T0237652

Water Resources Development and Management

Series Editors

Asit K. Biswas, Singapore, Republic of Singapore
Cecilia Tortajada, Singapore, Republic of Singapore

Editorial Board

Dogan Altinbilek, Ankara, Turkey
Francisco González-Gómez, Granada, Spain
Chennat Gopalakrishnan, Honolulu, USA
James Horne, Canberra, Australia
David J. Molden, Kathmandu, Nepal
Olli Varis, Helsinki, Finland

More information about this series at http://www.springer.com/series/7009

Yahua Wang

Assessing Water Rights in China

 Springer

Yahua Wang
School of Public Policy and Management
Tsinghua University
Beijing, China

ISSN 1614-810X   ISSN 2198-316X   (electronic)
Water Resources Development and Management
ISBN 978-981-13-5314-7   ISBN 978-981-10-5083-1   (eBook)
DOI 10.1007/978-981-10-5083-1

© Springer Nature Singapore Pte Ltd. 2018
Softcover re-print of the Hardcover 1st edition 2018
This work is subject to copyright. All rights are reserved by the Publisher, whether the whole or part of
the material is concerned, specifically the rights of translation, reprinting, reuse of illustrations,
recitation, broadcasting, reproduction on microfilms or in any other physical way, and transmission
or information storage and retrieval, electronic adaptation, computer software, or by similar or
dissimilar methodology now known or hereafter developed.
The use of general descriptive names, registered names, trademarks, service marks, etc. in this
publication does not imply, even in the absence of a specific statement, that such names are exempt
from the relevant protective laws and regulations and therefore free for general use.
The publisher, the authors and the editors are safe to assume that the advice and information in this
book are believed to be true and accurate at the date of publication. Neither the publisher nor the
authors or the editors give a warranty, express or implied, with respect to the material contained
herein or for any errors or omissions that may have been made. The publisher remains neutral with
regard to jurisdictional claims in published maps and institutional affiliations.

Printed on acid-free paper

This Springer imprint is published by Springer Nature
The registered company is Springer Nature Singapore Pte Ltd.
The registered company address is: 152 Beach Road, #21-01/04 Gateway East, Singapore 189721,
Singapore

Preface

This book has been a long time in gestation. It originated from my PhD thesis submitted to Tsinghua University in 2004. The manuscript entitled "水权解释" (shuiquan jieshi) was then published in 2005 by Shanghai Sanlian Book Store and Shanghai People's Publishing House. The literal translation of this title is "Explaining Water Rights". My pursuit of researching into the topic "water rights" is traced back to the year 1998, when flow cut-off of Yellow River raised grave concern in China. At that time, I was about to be a graduate student at the Research Institute of Twenty-first Century Development at Tsinghua University and involved in Professor Angang Hu's research project around "Yellow River Basin Management". Professor Hu and I presented an argument that directive allocation could not solve problems of water resources allocation in China and "quasi-market" mechanisms are critical to Chinese water resources management. In 2000, our paper "Public Policy of Water Resources Allocation in the Transition: Quasi-Market and Democratic and Consultative Politics" was published in a Chinese academic journal, which received positive responses and led to a public discussion of issues of water rights and water market.

The opportunity of one-year fieldwork from 2002 to 2003, as assistant director of Water Administration Bureau of Yellow River Conservancy Commission, provided me with concrete evidence and a thorough understanding of water resources management in Yellow River Basin. During my field trip in Shaanxi province, I was filled with excitement at the thought of "hierarchical governance structure" in explaining the link between water governance and state governance. The concept "hierarchical governance structure" suggests an alternative approach to studying water resources management in China. The hierarchical water governance structure in China originated in the Qin Dynasty that unified the country 2000 years ago, and such a unique structure has been continued till the present without interruption despite changes of rulers. On this basis, "water rights hierarchy conceptual framework" is developed to analyse issues of water resources reform in contemporary China. It constructs a choice model for water governance structure and advances the logic for structural choice with the minimum transaction costs under constraint conditions, which is demonstrated from the perspective of both management cost and cooperation cost as the inherent nature of China's choice for such a hierarchical structure.

During this process, several papers and research reports were written in an attempt to elucidate the picture of Chinese water resources reform: Whether water rights and water market are appropriate to China? How to introduce and implement such a reform? My academic writings had attracted considerable attention in China from both academia and practitioners, and they provoked heated debate among practitioners in Water Resources Ministry and Bureaus. Since then, I have been one of the leading scholars in the domain of Chinese water governance, water rights and water market. This book aims at contributing to the basic theory of water rights, providing a theoretical basis to assist policymakers in achieving in-depth understanding of water rights in the ongoing water resources reform in China. My argument is developed on three key theoretical standpoints: Cheung SNS's "economic explanation", Douglas C. North's "new economic history" and Ray Huang's "macro history". It gives a systematic explanation of changes of the water rights systems in ancient, modern and contemporary China, elucidating the inherent mechanisms in water rights allocation and exploring the possibility of introducing a dynamic market mechanism into the hierarchical governance structure.

This book is a product of the Major Program of the National Social Sciences Foundation of China (15ZDB164) and is jointly supported by the National Natural Science Foundation of China (71573151), Tsinghua University Initiative Scientific Research Program (2014z04083), the China Institute for Rural Studies at Tsinghua University (CIRS2016-6) and the National Key Research and Development Program of China (2016YFC0401408). The book has been benefited greatly from the support of many people, who I am deeply indebted to.

Mr. Shucheng Wang, the former Minister of Water Resources, has provided me with generous support during my fieldwork and internship in the Water Administration Bureau of Yellow River Conservancy Commission. In addition, working with the colleagues in this bureau has significantly facilitated and enriched my fieldwork in Yellow River Basin management, due to their personal connections with research participants and their knowledgeable conversations with me.

I owe an immense intellectual debt to my PhD supervisor Professor Angang Hu, who was an academic father figure to me throughout the journey of researching and drafting this book. In 2009, I was fortunate enough to obtain a scholarship to do research with Professor Elinor Ostrom at Indiana University, who was awarded the Nobel Prize in economics in the same year. She read this work and made critical comment. She has been the inspiration for my academic pursuit in the domain of governing the commons, and the experience of working with her has shaped my approach to researching into the role of institutions in commons. Without Professor Asit K. Biswas and Dr. Cecilia Tortajada's helpful comment and recommendation, this book would not have been published in this book series of Water Resources Development and Management. Following their helpful suggestions, this book has been updated with further research and understanding of water rights development in China, on the basis of the Chinese publication in 2005. I appreciate the excellent editing from Dr. Tingting Wan, as well as the assistance from Ms. Heyin Chen, before the book was submitted to the publisher. I also would like to thank the book production project coordinator, Mr. Prasanna Kumar N., and the editing team for their patience, communication and input during the publishing process. Finally, I thank my family for their endless support and love along my long academic journey; the birth of my little boy, Johnny, brought great joy to me during the past year, and this new book is dedicated to him as a special gift.

Beijing, China   Yahua Wang

Contents

List of Figures

List of Tables

Chapter 1
Introduction

1.1 Research Background

1.1.1 Context of Water Governance in China

Water governance is of extreme importance in China. The tradition of water governance is as long as the history of the Chinese civilization. As early as in the Spring and Autumn period (770–476 B.C.), a noted statesman Guang Zong stated that a good ruler must first of all eliminate disasters caused by floods and droughts (Gu, 1997). History of this big agricultural country of China shows that accomplished rulers in China, such as the first emperor of the Qin Dynasty, Emperor Wu of the Han Dynasty, the First Emperor of the Tang Dynasty and Emperor Kang Xi of the Qing Dynasty, all devoted full energy to water governance. All the heydays in Chinese history are associated with the achievements in water governance. When irrigation develops well, the country would get stable; with political stability, the people would feel secure; when the people feel secure, they would have the zeal for production; when there is enough food, all other trades and services would flourish and the whole society is bound to prosperity so no foreign enemy would dare to invade. Otherwise, there would be such severe disasters as floods and droughts that would throw the economy into recession and the people into misery, driving people to rise in revolt, thus plunging the country into turmoil and forcing the country to change ruler, even if there is no foreign invasion. Thus, eliminating floods and building irrigation projects is a matter of great importance in making the country prosper and pacifying the neighbors.

The vital position of water governance in China is determined by the specific national conditions. Due to differences in topography and the monsoon climate, precipitation in China varies greatly in time and space. In most places, the 4-month rainfall in a year makes up 70% of the annual total, resulting in an extremely uneven distribution of water resources during different times of the year and among different regions. North China is short of water resources. Though the land area

© Springer Nature Singapore Pte Ltd. 2018
Y. Wang, *Assessing Water Rights in China*, Water Resources Development and Management, DOI 10.1007/978-981-10-5083-1_1

in the north of the Yangtze River makes up 64% of the national total, its water resources account for only 19% of the national total. The distribution of river flow among different years and even within the same year is extremely uneven. The annual maximum and minimum runoff ratio may reach as high as more than ten times in North China. The water distribution within a year is mainly concentrated in the flood season when the runoff accounts for more than 80% in the North China Plain. The annual change of water resources is so big, its distribution within a year is so concentrated and the changes in rainy and dry seasons are so capricious that they are rarely seen in the other parts of the world. And that is the reason for frequent flood and drought disasters. According historical records, in the 2155 years from 206 B.C. to 1949, there were 1092 big floods and 1056 droughts in China, almost happening every year (Gu, 1997, pp. 229, 233).

Although Chinese rulers in the long history had the political desire to control water, none succeeded in bringing floods under total control due to limitation in technology, financial situation and governance abilities. Rather, the frequency of flood and drought increased with the passing of the time. According to historical records, the annual occurrence of flood and drought disasters was 0.6 times in the Sui Dynasty (581–618), 1.6 times in the Tang Dynasty (618–907), 1.8 times in the Northern and Southern Song dynasties (960–1279), 3.2 times in the Yuan Dynasty (1279–1368), 3.7 times in the Ming Dynasty (1368–1644) and 3.8 times in the Qing Dynasty (1644–1911) (Hu, 2003, p. 334).

The dyke of the Yellow River, known as the "sorrow" of the Chinese nation, breached every 26 years in the Qin and Han dynasties. However, when it came to the Yuan, Ming and Qing dynasties, it breached every 4 to 7 months. Over the past more than 60 years since the founding of New China, the country, under the leadership of Chinese government, has not only succeeded in containing the threat by floods and droughts and making the Yellow River safer, but also ensured adequate food and clothing of the 20% of the world's population with only 6% of the world's renewable water resources, an achievement that has no precedence in the Chinese history. Since the country introduced the Reform and Opening-Up Policy in 1978, great changes have taken place in water governance, resulting in new problems such as shortage of water resources and water pollution. Compared with all past dynasties, water governance requires an entirely new implication in the new period of development, posing new challenges to the state in water governance.

1.1.2   Water Crises in the New Period of Development

Since the Reform and Opening-Up Policy, China has witnessed unprecedented social changes and rapid economic development. People's living standards have improved significantly. However, the dramatic progress in urbanization and industrialization has sharpened the conflicts between population and resources dramatically. The ecology experiences deteriorating extensively; and environmental pollution becomes very serious. These have resulted in a resources crisis and ecological deficit in a dimension that is unprecedented in history. In this context, water issue comes to prominence. The general situation of water resources has become extraordinarily grave and complicated, especially the problems of floods, water shortage and deterioration of ecological environment. The 1998 floods of the Yangtze River, dry-up in sections of the Yellow River in the 1990s, pollution of the Huaihe River and dry-up and pollution of all the streams in the Haihe River catchment area are the typical cases concerning the three major problems.

Through half a century's water control efforts, the country has built dykes to protect major rivers, forming a flood prevention network with a web of reservoirs and flood retention areas, and brought the floods of some small rivers under control to varying degrees. Yet, the flood prevention projects of most rivers are not high in standards, capable of controlling floods that only happen in 20 to 50 years but not for bigger floods. The 1998 floods caused a direct economic loss of 300.7 billion yuan or 3.8% of the GDP of the year. With the increase of population and development of the economy, the losses caused by floods would rise drastically. This has raised a higher demand on the security of flood prevention projects.

Shortage of water has become more and more acute. At present, water resource per capita in China is only 2200 m³, hardly a quarter of the world's average. The figure for the Haihe, Huaihe and Yellow River basins ranges from only 350 to 750 m³. Furthermore, serious water pollution and water loss and soil erosion have made things even worse. Now more than 400 of the 669 cities in China do not have adequate water supply and 110 others are in acute water shortage that amounts to 6 billion cubic meters a year and dents the industrial output value by more than 200 billion yuan. The annual water shortage in agriculture has run up to 30 billion cubic meters and many rural areas have not enough water-associated sanitary facilities. Water shortage has become a major factor that holds up economic and social development, especially in the northern part of the country.

The water ecological environment has been fast deteriorating. Nearly 50% of water sections and the water surfaces of 90% of the cities have been polluted to varying degrees. Most places in Northwest, North and Middle China suffer from water shortages to varying degrees. The situation has been aggravated due to serious water loss, soil erosion and serious water pollution, knocking the water ecology off balance, causing sections of drying up of rivers, lakes and wetlands, invasion of seawater to land, desertification, and degeneration of forests and grassland.

China has the world's largest population and its water ecological system is fragile. Looking into the future, with the further economic and social development, water resources, water environment and water ecology will become more and more insecure that is likely to become one of the major crises threatening the survival and development of the Chinese nation in the next decades.

In the first half of the current century, the increasing population will make water shortages more acute. When China's population reaches about 1.6 billion by 2030, the per capita water resources available would drop from the current 2200 m^3 to about 1700 m^3, close to the warning line commonly recognized in the world. That will make the water shortage in parts of the country even worse. The acceleration of urbanization will lead to water supply crisis in cities and towns. China's city population was 220 million in 1980 and the water consumption was 16.8 billion cubic meters. The city population reached 306 million in 2000 and the water consumption reached 16.8 billion cubic meters. In the future, the rural surplus labor would move into cities at an annual rate of more than 10 million. It is predicted that the national economy would require additional 140 billion cubic meters water by 2030 mainly for the consumption by industry and urban population. Wastewater discharge will also increase accordingly that is likely to increase from 58.4 billion tons in 1997 to 85 to 106 billion tons by 2030 (PLDMWR, 2002a). The increase in the water consumption in production and daily life will deplete water required in the ecological environment. In addition to the further increase in wastewater discharge, this will make the fragile water ecological environment worse.

It is expected that water will become the biggest resources bottleneck limiting Chinese economic and social development in the next decades when the country is striving to edge into the high-income countries in the world. Water issues will pose the biggest challenge to human development in China. Water crisis will remain a hidden danger in national security. How to ensure water security, including security of water resources, water environment and water ecology and bring about a sustainable use of water resources—this has become one of the most important issues of development in the current century.

1.1.3 Changes of Governance in Response to Water Crises

Superficially, water crisis is a resource crisis. However, in essence, it is a crisis of governance. It is the ultimate upshot of the long-term lag of the water governance system. Water crisis reflects the fact that water governance system has for long been unable to respond to the changing social conditions. The concept is outmoded; policy adjustment is slow; institutional construction is lagging behind; water control abilities are so poor that they fail to match the increasingly complicated water governance environment, being unable either to optimize the inadequate water resources and ease the sharpening conflict between water supply and demand or coordinate the steps between upstream and downstream, among different regions and departments. The fundamental resolution lies in the change of ways of how water resources are governed and how to seek a new water governance model to replace the old one under the conditions of the market economy, globalization and widespread application of information technology.

The Chinese government intentionally accelerated the revolution in the way of water resources governance toward the end of the 1990s when it was seeking countermeasures against the complicated water problem, especially in the course of preparing plans of action with regard to the control of the Yangtze River floods, the flow cut-off in sections of the Yellow River and the diversion of water from the south to the north of the country. This revolution, marked by the series of important speeches by Wang Shucheng, the former minister of Ministry of Water Resources has achieved attractive theoretical development, including the proposition of transition from the engineering-oriented model to the resources-oriented model, from traditional model to modern and sustainable model, and while continuing to prevent floods and other natural disasters and mitigate their impact, putting in the first place the solution of water resources shortage and water pollution so as to support the sustainable economic and social development with the sustainable utilization of water resources. This new line of thought in water governance can be summed up in the following three aspects: (1) new water governance principle: harmonious co-existence between human and nature; (2) new management system: integrated management of water resources; and (3) new instruments of water governance: water rights and water market. The following is a brief account of these three aspects (Wang, 2002).

First, change in the principle of water governance. The new water governance principle developed on the basis of the traditional one holds that water is the basic natural resource and a controlling factor in the ecological environment. In governing water, it is imperative to stick to the principle of acting upon the natural law and make man co-exist with nature in harmony instead of simply taking from nature in an unrestrained manner. While preventing human from being harmed by water disasters, special care should be taken to prevent human from harming water. While developing, utilizing and governing water resources, special attention should be given to the allocation, conservation and protection of water resources. It is necessary to pay attention to the close relations between the ecology and water and put the water needed by the ecology on the major item of agenda. It is essential to prevent the damages of the ecological environment by exhausting water resources. While paying attention to the building of water control projects, special attention should be paid to non-project measures and scientific management. It is also necessary to shift from supply-side management to demand-side management and work out master plans of the national economic development according to available water resources.

Second, exploration of management system. The new way of water governance highlights that water takes watershed as a unit, and surface water and ground water are mutually transforming into each other and the development and utilization of upstream and downstream water. Besides, two banks of a river and between mainstream and tributaries are mutually influencing each other; volume and quality of water are interdependent; and the development and utilization of water, including flood prevention, flood control, water retention, water supply, water use, water conservation, water discharge, sewage treatment and recycling of neutral water are all inter-related. These require unified arrangements and planning and comprehensive control in order to achieve scientific and rational allocation of water resources. The major defect of the traditional water management system is "multi-partite water governance". The new way of water governance requires unified management, unified planning and unified regulation of water resources in a watershed and active efforts to seek ways of unified management of surface water and ground water in cities and the countryside so as to gradually realize unified management of a watershed and integrated management of regional water affairs.

Third, innovation of water management means. According to the new way of water governance, water is a commodity and a strategic economic resource. The new market economy requires organic integration of macro control by the government and market mechanism according to the law governing the development of the economy, full display of the pulling effect of the market in the allocation of resources. It is, therefore, necessary to seek ways of building up a water rights system and water rights market, pushing the reform of water conservation financing system and pricing mechanism, encouraging economical use, optimal allocation and effective protection of water resources. The new way of water governance also gives much stress to the improvement of related laws and legal means and technical progress. It also stresses the necessity of widespread application of information technology in water management so as to stimulate and step up the pace of modernization in this area.

The new way of water governance, improved in practice and used to direct new practice of water governance, has achieved initial successes. The following examples are the outcomes born out of the new way of water governance. All of them mark a shift of the revolution in China's way of water governance from theoretical plan to field practice (Wang, 2003a). First, constructing river flood prevention system. The rare floods of the Yangtze River in 1998 have led to a change in the way of flood prevention, a shift from the disorderly and unrestrained scramble for land between human and water to orderly and sustainable harmony between human and floods. Starting from 1998, the Chinese government has invested tens of billions of yuan in a mammoth integrated flood prevention system for the Yangtze River, resettling nearly 2.42 million people displaced by the project of returning land reclaimed from rivers and lakes, restoring 2900 km^2 of water surface and increasing flood holding capability amounting to 13 billion cubic meters. In 2002, although floods occurred in the middle reaches of the Yangtze River and the Dongting Lake area, all of the other rivers and lakes around were in good condition.

Second, easing the flow cut-off in sections of the Yellow River. Yellow River flow was interrupted almost every year in the 1990s. The most serious one happened in 1997 when the river bed in some sections ran dry for 226 days. Starting from March 1999, the State Council authorized the Yellow River Conservancy Commission (YRCC) to exercise rational allocation and unified diversion and arrangements of water in the entire basin for production, living and ecology, encouraging planned and economical use of water. Up till now, no section of the river has run dry for successive years despite long dry spells, thus ensuring the water for living and production in riparian cities and rural areas. The ecology downstream has also improved. Without the unified management and regulation of water, it would have been hardly imaginable to maintain the flow of the river.

Third, developing the South-to-North Water Diversion Program. The program is an important and basic facility for optimizing the allocation of water resources and a move of strategic importance in easing water shortages and alleviating the deterioration of the ecological environment in North China. By following the principle of "water conservation before water diversion, pollution control before letting water in and environmental protection before using water", the Chinese government has carried out an all-round, in-depth and scientific feasibility studies of the water diversion project before coming out with a master plan. In order to ensure the anticipated result, the plan introduces the new mechanism of water right, water price and water market to put the project under the new management system of the "state exercising overall control, companies engaging in market operations and end-users participating in management". The water allocation has been implemented according to this new mechanism.

Fourth, irrigation projects in Western China Development Drive. Western China is short of water resources and its ecological environment is fragile. Water development is the foundation and key to the development of western China. China has accelerated the pace of building irrigation projects after 1999, with emphasis put on the restoration of the ecological system and environmental protection. Work has started on the comprehensive control of the Tarim River and the Heihe River basins to divert water to the upper reaches where the ecology has been seriously deteriorating. Water flow has come back in the 363 km downstream of the Tarim River, the largest inland waterway in China, rescuing the desert vegetation on the verge of extinction and injecting renewed vigor into the green corridor. The ecology downstream of the Heihe River, the second largest inland waterway, has improved significantly.

Fifth, implementing new water law. Starting from October 1, 2002, China began to implement the revised version of the *"Water Law of the People's Republic of China"*. Drawing on the new experiences and new principles prevailing in water governance in recent years, the revised version of the water law has codified the principles, line of thinking and objectives of water governance advanced in recent years. Compared with the old version, the new law has the following features: (1) reforming the water management system according to the principle of integrating basin management and administrative regional management, with stress laid on unified management and on the legal position of basin management organizations; (2) giving priority to the economical use of water and the raising of water utilization efficiency and strengthening of management of the use of water by controlling the total quantity, supplemented by quota management, and introducing the water license system and compensatory use system; (3) strengthening overall control of water resources and clarifying the legal status of water resources planning, specifying a series of legal systems for intensifying the management of the allocation of water resources; (4) better coordinating the relations between water resources and population and between economic development and ecological environment, stressing the protection of the ecological environment in the course of developing and utilizing water resources (PLDMWR, 2002b).

The above theories and practices concerning water governance show that China has started a revolution of far-reaching significance in water governance. It means not only a change in water governance model during the smooth transition from the planned economy to a market economy over the past dozens of years, but also a profound revolution in the traditional water governance that have prevailed for several thousand years. With the objective of raising utilization rate of water resources, optimizing the allocation of water and realizing the sustainability in the use of water, the revolution is guided by the thinking of changing from engineering-oriented governance model to resource-oriented governance model and of realizing the harmony between human and nature, introduces water rights and water market and other new water management means and promotes the reform of the water management system toward integrated basin management and integrated regional water management.

1.2 Origin of Water Right Issues in China

The grave water source situation, the unique national conditions and the transitional economy have determined the complexity, arduousness and long-term nature of the change in the ways of water governance. Gratifying progress as it is in the revolution of water governance, the revolution is still riddled with many theoretical and practical problems. One of the core problems is how to coordinate the relations between the government and the market under the condition of the new market economy.

As the market-oriented reform is going into depth, market mechanism is playing an increasingly basic role in the allocation of economic resources and it is penetrating into the allocation of all resources. For a long time, the Chinese government has played the leading role in water resources management. The market-oriented reform and the exacerbation of water shortages have increased the pressure on the introduction of market mechanism into water allocation. Can China introduce market mechanism to optimize the allocation of water under the conditions of the new market economy? Water is a basic resource of strategic importance and a controlling factor in the ecological environment. Can water resources be allocated by the market forces with clear definition of property rights like other economic resources? If so, to what extent will the market mechanism play its role in the allocation of this special resource? These touch off a great debate concerning water rights and water market (Wang, 2000).

The debate has unfolded around three main views. The first is that the market is the foundation. This view sees no fundamental differences between water resource and other economic resources and, therefore, water resource can be allocated by way of defining property rights or by way of property rights auction (Sheng, 2002).

The second is market skepticism. Though seeing that the introduction of market mechanism is inevitable, it doubts whether the conditions are ripe for doing so under the current legal framework and governance structure that water resource allocation relies on unified management and administrative means (Yang, 2002). The third is that market can only play limited roles. It holds that due to the special nature of water resource and the actual conditions in the transitional period, water market can only play limited roles. It can be termed as "quasi-market" or "incomplete market" (Hu & Wang, 2000; Wu, 2002).

The debate reached its zenith when the first water right transaction appeared in Zhejiang Province between Dongyang City and Yiwu City. The transaction really happened under the legal framework of public ownership of water resource and water right transfer is forbidden (Yang, 2002). What happened in Zhejiang entailed a number of case studies, bringing into depth the water rights and water market debate.[1] In the process, a number of international literature were translated into Chinese that opened the horizon of the people. Through extensive discussions for nearly 3 years, the debate cooled down with the basic common understanding that it is necessary to build a water market but it will take a long time to do so (Yahua Wang & Hu, 2007). On the one hand, in order to raise the water utilization rate and ease the conflicts between supply and demand and increase input in water projects, it is necessary to introduce the market mechanism; on the other hand, it would require a long period of time to build up and improve the water rights system and establish the water right turnover mechanism subjecting to the macro control by the government and conforming to market rules.

Rather than the necessity of introducing market mechanism, it is concerned with how to introduce it. What to be resolved in the first place are the problems concerning water rights allocation that are difficult both in terms of theory and practical operation. To whom will the state allocate the primary water rights? How to give equal consideration to fairness, efficiency and ecological environment in allocating the primary water rights? How to integrate water rights allocation with unified management so as to achieve real sustainable use of water resources? Although there are some studies around these issues, there are no convincing answers yet. This reflects that theoretical studies of water rights are lagging behind. This book aims to answer questions concerning water rights allocation in a order to unveil the inherent mechanism of how to define clearly water rights so as to provide theoretical basis for building up the water rights system and references for developing a water market in China.

---

[1]The Ministry of Water Resources called a number of symposiums centering on the subject. Articles have appeared by the hundreds, which were collected in two volumes of a book entitled "Water Rights, Water Price and Water Market Forum" by the Policy and Law Department of the Ministry of Water Resources in 2001. The China water information website http://www.waterinfo. net.cn opened a special forum in this subject for uploading related information and articles. This resulted in the publication of a number of "Collections of Articles" or books on water rights and water market.

From the long-term point of view, China's system of water allocation will be a mixed one that will integrate planning mechanism with market mechanism. Although we are not clear at this stage about how to integrate administrative means with market means theoretically. Which means should be used more, which should be used less, and how to adjust the water management operational mechnism after the introduction of market mechanisms? This book also attempts to answer these questions on the basis of basic theoretical study. These questions are, in fact, central to the changes in water governance, that is, how to coordinate the relations between the government and the market in water allocation under the new historical conditions, and how to coordinate the relations between the government and the market in water governance.

1.3   Theory and Methodology

1.3.1   Literature Review

The term "water rights" is, in fact, entirely new in contemporary China. As water right is a very complicated issue, there is even divergence of views on the basic meaning of this term. The studies of water right theories are just in the initial stage. However, there have been some achievements in the basic theories in the academic circles in recent years. They can be divided into three major categories in terms of methodology. One is the study from the angle of law as represented by Shouqiu Cai (2001), who studies the legal principles relating to water right transfer. Jianyuan Cui (2001) studies the basic theoretical problems of water rights from the angle of civil law, and Bin Liu (2002) analyzes the legal framework for the water rights system in China. Another is the study by Wenlai Jiang (2001) using economics methodology to analyze the modern Western property right theories which are regarded as the foundation of the theories of water rights. Yunkun Chang (2001) examines the historical changes of the water rights system of the Yellow River from the perspective of institutional economics. Hong Sheng (2002) gives an in-depth analysis of the water rights system and water market by using the methodology of classical economics. Games theory to study different types of allocations of water rights is discussed (Hu, Fu, & Wang, 2003). The third is the international comparative studies to introduce foreign experiences for reference in the building of China's water rights system (Calow, Howarth, & Wang, 2009; Sun, Wang, Huang, & Li, 2016), which offer useful experiences of developing water rights in China based on foreign evidence in the field of water rights transaction. Comparative studies of the water right systems and water markets are conducted, which make propositions and recommendations for improving China's water management system.

Water market in Western countries has a history of more than 30 years. These countries have accumulated a wealth of experience and lessons for China to learn. For a country as China, who is exploring for a water right system like "wading across a river by feeling the stones" as a Chinese saying goes, it is no doubt a low cost option to learn from foreign experience would be beneficial. So far there is a rich bank of translated materials about foreign water rights and water markets, providing a good basis for comparative studies. However, as the water management system of a country is closely associated with its political and economic systems and the level of social and economic development, the ready-made practice and experience can serve only as a reference for us and cannot be aped. China has very unique political and economic structures and therefore needs to seek a path best suited to its own conditions. These impose the limitations of applying foreign experience and require us to enhance theoretical studies so as to get a better understanding of the Chinese case while studying foreign experience.

However, in terms of practical results, it is believed that law has a greater impact on the policies and practice. The basic viewpoint on water rights from the angle of law is the "separation of ownership and use right". Legal experts point out that "the right to use water resource is not the right to the property, but an independent right derived but different from the right to water as an asset; water ownership is a kind of absolute right to property and the water use right is a usufruct; and the owners of water rights can authorize the use right to non-title holders, thus separating water ownership from water use right" (Cai, 2001, p. 297). Such separation has become the basic understanding of water policymakers on water rights. Water rights include ownership and use right. Since the Chinese law provides that the ownership of water resources belongs to the state, what we discuss in reality refers to the use right. The allocation of water rights is the allocation of water use right (PLDMWR, 2002a).

Based on the thought of "separating ownership from use right", there are some operating plans for allocating water rights and introducing water market. For instance, "application should be filed with the state for obtaining the right to extract water or the right to use state-owned water resources. All people desirous to use state owned water resources must file applications for the water use right and obtain permits. The water rights holders may transfer all or part of the water rights held to prospective appropriators. The obtaining of the water rights is the primary market, monopolized by the state; water rights transfer is the secondary market, operated under the macro control by the government, which exercises regulation and standard management of the conditions, procedures and ways of water rights transfer and water property right transaction market" (Zhao, 2001, p. 49).

The above views reflect the basic understanding of water rights issues from the angle of law. The proposition of separating ownership from use right could easily lead to plans of execution with regard to law, regulations and policies. However, such a study does not gain thorough understanding of water rights definition and allocation associated problems. For instance, how to look at the relations between use right and ownership right? Under what conditions is it feasible to turn the water withdrawal right endowed by the administrative permit into the use right that puts property rights to beneficial and non-wasteful use? By what degree can the use right be turned over through the market? These and other problems cannot get clear-cut answers from the angle of law. In addition, the actual operability of policy plans designed from the angle of law is subject to doubt.

The economics methodology, is most important to fill the gap left by the other two categories of studies. Economics methodology focuses on the inherent mechanism for the operation and development of the subject studies and reveals the inner ties among objective things. Through the interpretation of complicated problems in reality, it serves the purpose of implementing policies and laws. Some Chinese economists have already achieved some successes in the difficult subject of water rights allocation. For instance, games theory is used to study water right allocation that advances the idea of adopting administrative means in the primary allocation of water for consumption and applying market rules in reallocation (Hu et al., 2003, pp. 84–113). Market means are applied to regulate water resources, including auction mechanisms for primary water right allocation and water right trading (Sheng, 2002). Chang Yunkun studied the evolution of the water right system of the Yellow River and called for a clear definition of the water rights, fixing the water prices and cultivating the Yellow River water market and designed rules for water right trading (Chang, 2001, pp. 179–191). Perhaps due to professional bias, some economists always go overboard for studying the market allocation of water resources. However, such conclusions as it is necessary to introduce the market mechanism into the allocation of water resources fall far short of the demand for knowledge about water rights. In designing an ideal water market allocation plan, they are in fact attempting at something they are not good at. In the academic division and specialization of labor, the power of economics lies in the interpretation of the actual world, but not criticism of the real world, let alone designing an ideal world. Economics methodology makes more basic contributions to the allocation of water resources and provide the theoretical support for the implementation of laws and policies by enhancing the understanding of water rights.

1.3.2 Principles and Methodology

The book attempts to contribute to the basic theories on water rights and presents its views on the water rights in contemporary China. In general, the water rights studies in China are mainly application-oriented, with more principled discussions focusing on water rights allocation and water market and policy recommendations concerning the building of the water rights system and the rules on water market. There are little in-depth studies of the basic theories that go deep into the essence of water rights, inherent mechanism for a clear definition of water rights, the logic for the evolution of water rights and the laws governing the development of water market. Academic study has a very strong nature of continuity. The current conditions in the water rights theory study indicate that it is an area urgently in need of development. When an area of study is still in the primary stage of development, people would, often than not, come out with remedies to practical problems based on perceptive knowledge. It is, of course, necessary to timely meet the requirements for the implementation of policies. But with the revelation of underlying conflicts in the course of policy implementation, policymakers would no longer be satisfied with the superficial interpretations and therefore raise higher demand on basic theoretical studies. China's water rights studies are exactly in this stage. This book may, therefore, be regarded as a response to this social demand.

This book has adopted the economics methodology. It follows the tradition of neoinstitutional economics and uses its methodology to study China's water rights issue. The fundamental purpose of academic studies is to enhance the people's understanding of the real world instead of airing views and recommendations based on the differences between the real world and ideal world by proceeding from the "virtual ideal world". Economics is the most mature discipline in social sciences after more than 200 years of development. The analytical tools it has developed have become more and more instrumental to interpreting the realities and it is truer with the neoinstitutional economics. Modern economics is able to explain logically most of the economic phenomena observed. Economics studies may get its strength from its explanations of the real world instead of winning approval for discovering inconformity between reality and theory. The basic principle and starting point of this research is to study the water rights structure in the real world by proceeding from the de facto objective rights relations rather than proceeding from articles of law. The methodology of institutional economics adopted in water rights issue is a kind of property right relationship and its essence is the people-to-people institutional arrangements. Institutional economics is the most appropriate method for studying this issue.

This study attempts to discuss problems within a logically viable theoretical framework made up of a group of interconnected models and theoretical propositions that is generally called "hierarchy theory". The real world is always very complicated. However, by relying on theoretical models, we have simplified the understanding of the real world, as scientific models may provide an insight into the laws governing the development of things and play a guiding role in human behavior (including policy implementation). Usually, there may be a number of models to choose for a given economic phenomenon. Which model to choose depends on the cost (complicatedness of model) and benefits (power of interpreting the real world) of the use of the model. Although it is not always necessarily so, the interpretative power of models is likely to rise with the increase in precision and will soon go beyond the comprehension of most people. Academic development will become more and more detailed in the division of labor. The use of "hierarchy model" as a theoretical tool to describe water rights relations is an effort to uplift the professional level in this area, as it is the most interpretative model among all options available. Water rights theory is a promising academic area and the professional level is comparatively low. For the current development level, conceptual model is the most suitable in order to get a general knowledge of the issue. With the knowledge going deeper and deeper, the division of labor will become more and more minute and the subject of study will become more and more refined and more and more mathematical and physics models will be applied. This book has adopted conceptual model in many places as this is the best cost-effective option for the current professional level.

Academic contribution is always marginal. This book attempts to use the latest achievements, especially those in the social sciences of the West and takes a small step forward on "the shoulders of the giant". As far as water rights theory is concerned, it is a medium-level theory covering a very small area. Water rights issue is but one of the problems concerning the property right theories of natural resources in the western modern economics development. The water rights theory per se is not an independent discipline. It seems that water rights problem does not have a strong particularity in the western society. The general property right theory and further theories of property right of natural resources are enough to meet the practical demand. China needs specialized water rights theory in that water rights problem in China who has its strong particularities that are not only different from Western countries but also in urgent practical demand in its economic and social transitional period, in which many of the underlying problems call for answers from the theoretical plane. That is why it is extremely necessary to discuss theories at this medium level in China. The medium-level theory may be built on the achievements at a higher level. So the western modern property right school and natural resources property right theories constitute the foundation for the study of water rights theories by this book. The hierarchy theory advanced by the book embodies the wisdom of many forerunning institutionalists. The core of hierarchy theory may trace back to Ronald Coase, the founder of new institutional economics. His theory about the nature of the firm is the earliest source of hierarchy. The hierarchy model of water rights structure comes directly from the theory "institutional hierarchy of natural resources" developed by Ray Challen, representative of the contemporary natural resources institutional economics, and Challen was, to a great extent, enlightened by Elinor Ostrom, a pioneer of the natural resources institutional economics who developed the "nested rules system", an embryo form of the institutional hierarchy. This model of institutional hierarchy advances the understanding on property rights of natural resources, in particular, for resources as water which is not static. This work has inspired succeeding studies of natural resources (Costa, 2015; Dou & Wang, 2017; Lukasiewicz & Dare, 2016; Marshall, 2013; McCann, 2004, 2013; Philpot, Hipel, & Johnson, 2016; Z. Wang, Zhu, & Zheng, 2015).

The book has devoted a substantial part to the empirical study of the hierarchy theory. Empirical significance lies, first of all, in testing the theory and observing the theoretical model to see whether it is the effective method for explaining the world and whether the theoretical model is interpretative to the real world. But the empirical study will never be adequate. It can verify the effectiveness of a theoretical model, but it cannot prove the correctness of a theory. The essence of science shows that no model is absolutely correct, because the essence of scientific nature is "falsifiability" and the unfalsifiable theory cannot be incorporated into the science domain. Any theoretical model is subject to modification and improvement, thus increasing its explanatory power. That is scientific method and spirit and in this process academic progress is achieved. Although the book has used a lot of empirical materials, such empirical analysis of the hierarchy theory is far from being adequate and it is, therefore, subject to modification and improvements by others and that process will enable the "water rights theory" to develop. Another implication of the empirical study is that in the process of applying the theory to the real world and meaningful proposition is derived. The part of empirical study in the book is instrumental to enhancing the understanding of the structural changes in China's water rights structure and supporting the proposals and recommendations for the reform of the water management system toward the end of the book.

This study has both a sense of history and a global perspective. It stresses the disparities between China and Western countries and this is useful in seeking a water right system best suited to the Chinese national conditions. The book stresses the particularities of the Chinese civilization, which are inherently associated with water governance. In order to have a better understanding of the water right problem in China as a whole, it is necessary to broaden the vision of history and review the relations between the Chinese civilization and water governance from a broader scope and get a firm grip of the continuity of water governance in the Chinese history. For this purpose, the book not only discusses the inherent connections between water governance and the origin of the Chinese civilization but also devotes a whole chapter to the study of the structural change of the ancient water governance structure in China. This has been proved to be very important to sustain the conclusions arrived at by the study.

1.4 Organizational Structure and Method of Narration

This book consists of eight chapters. Chapter 1 is an introduction. Chapters 2, 3, and 4 are devoted to theoretical discussions. From Chap. 5 to Chap. 7, empirical analyses are provided. Chapter 8 summarizes the book with conclusions and implications.

The introduction raises research questions and introduces the line of thought. This chapter introduces the background for the study: the contemporary China is in for a period of transition in water governance model that was handed down from the history of thousands of years as motivated by the new challenges of water shortage and pollution. But the transition in state governance has made the issue extraordinarily complicated. The core issue in the transition of water governance is the relations between the government and market, that is, how to display the role of market in water governance under the condition of market economy. The introduction of market mechanism in water management is key to the reform of the water rights system. But this reform is riddled with "difficulty in the allocation of water rights", that is, how to make initial allocation of water rights within the framework that the state holds the water ownership right. Then, the book comments on the three kinds of study methods and prevailing views on these problems, pointing out that the lack of basic theories has limited the understanding of the essence of water rights problem. This chapter also introduces the methodology and line of thought used in this research and its organizational structure. At the end, this chapter explains and defines some of the major terms and phrases frequently used in the book.

Chapter 2 is devoted to the study of China's water governance structure. The chapter expounds on the basic implications of water governance, pointing out that the most important output of water governance is to ensure water security and the collective action unfolded around it has constitute a continuum of governance structure. China's water governance structure is at the top of the hierarchy in the continuum. By introducing an operational model of governance structure, this chapter traces the earliest origin of China's hierarchical water governance structure. With very special natural geography, China used to be riddled with floods in the very early period of the Chinese civilization and that required a large scaled collective action, hence the need of a centralized social organization. The demand for centralization of power in water governance led to a unitary political structure in the early period. This is the water governance school of thought (Curtis, 2009; R. Huang, 2002; Needham, 1981; Wittfogel, 1957). The exposition of the relations between water governance and state governance echoes the viewpoints of this school of thought.

Chapter 3 presents the methodology for studying China's water rights structure. The property rights structure of water resources is very complicated. It does not only greatly differ from ordinary economic assets but also is the most complicated among natural resources. This chapter reviews the veins of development of the theories about the property rights structure of natural resources, focusing on discussion about the institutional hierarchy conceptual model. On this basis, it is modified into water rights hierarchy conceptual model to analyze Chinese water rights structure. In a hierarchical society, the water rights structure is also hierarchical. In the structure, the public power enjoys absolute precedence over the private power and interests, while private property right is in a disadvantageous position. At the end, the chapter compares the differences of the water rights structure between China and the West.

Chapter 4 presents the economics theory concerning water rights structure. The chapter reviews the development of the modern property right theories of the West, stresses the fundamental tasks of modern property rights analysis and uses economic explanation to unveil the rationality of the real world. By using the transaction cost analysis method, this chapter, first of all, explains statically why water rights assume the hierarchy structure and why a diverse of water rights allocation systems exists in reality. Then, it goes on with an dynamic explanation of why water rights structure evolves with the passing of time and what is the path option of the changes. The key to understanding this is the minimization of the constrained transaction costs. Static analysis takes only static transaction costs into consideration and in reality it tends to opt for institutional arrangements with the minimum costs under constrained conditions. Dynamic analysis introduces dynamic transaction costs in order to know why the inefficient institutional arrangements exist and the features of path dependence in institutional changes. The chapter devotes substantial spaces to the discussion of the motivation mechanism for introducing market into the hierarchy structure. For this purpose, the chapter introduces the concept of water rights quality as an important theoretical approach to understand the mutually replaceable relationship between administration and market. It also advances the view that the hierarchy structure excludes market mechanism and explores the motivation mechanism and possibilities of introducing market elements into the hierarchy system.

Chapter 5 is the empirical study of the structural changes of the water rights structure of ancient China. Taking the Yellow River basin as a site for empirical studies, the chapter describes the basic trend of the structural changes of water rights in the three historical periods of Qin-Han, Tang-Song and Ming-Qing dynasties by employing the water rights hierarchical conceptual model. Due to lack of water control projects, the ancient water rights problem mainly focused on the allocation of water for irrigation between communities and individual users. The book discovers that the Qin-Han was a period seeing the emergence of large scaled construction of irrigation projects and the irrigation management system. So the emphasis of study is placed on the water rights structural changes during the Tang-Song and Ming-Qing periods, which shows the following significant trends of change: first, the state gradually gave up the management of micro irrigation affairs and began to shift to macro control during the Tang through the Qing periods; second, resources quotas were used to divide up the water rights in the Ming-Qing periods; third, the water rights allocation shifted from application to registration in the Tang-Song through the Ming-Qing periods; fourth, transaction of water rights for irrigation occurred extensively during the middle and late periods of the Qing Dynasty. This chapter has given a detailed economics explanation of the historical phenomena by using the hierarchy theory advanced in the preceding chapters.

Chapter 6 covers the empirical studies of the water rights structural changes in modern China, i.e. since the founding of P.R. China. Still with the Yellow River basin as the site for empirical studies, the chapter uses the 'water rights hierarchical conceptual model' to describe the evolution of the water rights structure in the use of the surface water of the Yellow River basin and the evolution of the mechanism of allocating the rights to water for use in irrigation during the planned economy period and the transitional economy period. Then, it uses the hierarchy theory to explain the following three features of the evolution of the entitlement system: first, the replacement of the resources quota in the allocation of the right to water for irrigation by input quota since the country introduced reforms; second, the significant disparities in the water entitlement systems for irrigation in the upper and lower reaches of the river; and third, resources quota system has been universally introduced into other levels of the water rights allocation structure. The chapter also gives a detailed study of the operations of the current water rights structure, all citing the Yellow River basin, with emphasis put on the testing of the effectiveness of the water rights allocation regime at the sub-basin, local and group levels by using the data available about the Yellow River. The chapter arrives at the conclusion that the current administrative allocation regime is effective.

Chapter 7 focuses on case studies of the water market in contemporary China. This chapter dissects six cases that have happened over the past five years and divides them into five categories by applying the 'water rights hierarchy conceptual model': water rights transaction at the user level; water rights transaction at the group level; water rights transaction at the local level; water bank under the regulation of the superior authorities; and market allocation of initial water rights. Study shows that the introduction of market into the re-allocation of water rights falls in entirely with the hierarchy theory, that is, the introduction of market mechanism has become feasible because the water rights transactions are conducted in an environment in which the cost of administrative regulation is quite high and it happens that there are factors that can greatly lower the cost of market regulation. The case study also reveals the dilemma of the hierarchy water rights structure: with the changes in the market-oriented reform and other external environments, the management cost for maintaining the hierarchy governance structure rises and, due to inadequate input in management cost, there exists serious government absence in the allocation of water resources. In such circumstances, the introduction of market mechanism to a certain degree is a kind of cooperative cost paid by the society. This is a compensation for the inadequate input in the management cost by the government.

Chapter 8 is the conclusion of the study. The chapter sums up the most important ideas, views and innovations concluded by the study. First, it introduces the basic conclusions of the theoretical study and answers the three questions raised in the introduction: What is water rights? How to make the initial allocation of water rights? How to coordinate the relations between the government and the market in the water resources management under the market economic conditions? Then it comes the basic conclusions of the empirical studies, which include evolution of the water rights structure in ancient China; the evolution trend of the water rights structure since the founding of New China; and the prospects of the changes in water rights structure in the future. This is followed by some specific proposals and recommendations for promoting reform of China's water management system, which include changing the reform strategy from project construction focusing on institutional building and the specific policy recommendations concerning water management in nine categories. The book ends with the statement: The difficulty in water governance of contemporary China is, in essence, difficulty faced by the collective action in the contemporary social transition period and the challenge of water governance is, fundamentally, a challenge to state governance.

1.5 Definition of Key Terms

1.5.1 Water Resource

There are diverse views on the concept of water resources. The most commonly recognized definition in the academic circles is that given by UNESCO in 1997 that "'water resources' is water available, or being made available, for use in sufficient quantity and quality at a location over a period of time appropriate for an identifiable demand" (Yahua Wang, 2005). The commonly used term water resources usually refers to the renewable freshwater resources available on earth every year, including surface water and ground water. China's water law adopts this definition.

Compared with other natural resources, water resources are very complicated in its natural property and have a strong nature of public property: (1) Water resource is a "flowing" natural resource with watershed as a unit. Water, including surface water, ground water, water in top layer of soil, and vapor in the atmosphere, is in permanent motion, constantly changing from liquid to solid or gaseous phase, and back again. (2) Water resources have the "dual character of benefits and disasters". When water is less, there would be drought; when water is in excess, there would be floods. While building irrigation projects, people have to guard against floods. (3) Water resources take watershed as its unit, often spanning several administrative regions and that is easy to give rise to conflicts of interests among different regions in the development, utilization and governance of water resources. (4) Water resources are beneficial in many aspects of the economy, including water supply and irrigation, power generation, shipping, breeding and tourism. So the utilization of water resources concerns the balance of interests among different regions and among different departments. (5) The utilization of water resources involves water storage, water diversion, use of water and sewage discharge that concern the interests of all sides and it is very difficult to strike a balance. (6) Water resources are a controlling factor in the ecological environment. The utilization of water resources in excess of a certain amount would result in pollution, ecological destruction and other ecological disasters and it is very difficult to ease the conflicts between man and nature.

Water serves many purposes. Water resource, in its broad sense according to the way of benefits, usually includes water quantity resource, water energy resource and watershed resource. Some people say that it should also include water quality resource, which refers to the digesting pollution or self-purification ability and that is why it is also called water environment carrying capacity. In fact, the utilization of water quantity resources is closely related to consumption of water environment carrying capacity, water quantity and water quality. This book does not include water quality resource as it regards that water quantity resource itself includes water quality. Water energy resource can be used to generate electricity; watershed resource can be used for shipping, aquatic breeding and water-based tourism. Water energy resource and watershed resource are non-consumption water. Although they do have some impact on the water quantity resource and water quality resource, they can be relatively independent, especially they are mono-polistic in nature in the development and utilization. Their property right problem is much simpler. So they will not be discussed in this book. What the book studies is water quantity resource, which refers to freshwater resources for use in production and living, such as waterways, lakes, surface and ground water and water of arti-ficial rivers, lakes and reservoirs, ponds, pools, canals and other artificial water bodies.

Water resources are quite different from other ordinary economic resources in terms of allocation. The allocation of common economic resources pursues the maximization of wealth or benefits with limited economic resources. It is, in essence, a concept of efficiency value. The top pursuit of the allocation of water resources, however, is "water security", including security in potable water, in the prevention of floods, grain production, water for industrial use and ecological environment. The concern for security in the allocation of water resources is deter-mined by the particularity of the resource. Water is the basic natural resource, an irreplaceable daily life resource and a strategic economic resource and a resource that controls the ecology and the environment. Although it has the features of eco-nomic resources due to its rarity, the first objective pursued in the allocation of water resource is security. Efficiency is often subordinate to and serves the needs of security.

The allocation of water resources should ensure not only security but also equi-tability and acceptability. The basic and public nature of water resources deter-mines that it is easy to induce conflicts of interests. Only when the water resources allocation can ensure security, equitability and acceptability, is it possible to put the utilization of water resources to maximum efficiency. The subordinating position of efficiency in the allocation of water resources which determines that the allocation of water resources is far different from that of other economic resources.

1.5.2   Property Rights

In the first half of the last century, mainstream schools of economics discussed allocation of resources in an ideal world and expounded the effectiveness of the market mechanism in the allocation of resources under the assumption of zero transaction cost and the definition of property rights without conditions. Starting from Ronald Coase (Coase, 1937, 1960), the importance of property right in the allocation of resources has become more and more accepted. Coase spelt out the importance of the definition of property right and the rights and obligations arrangements in economic transactions in a number of his major theses, which was later termed as 'Coase Theorem' (Gjerdingen, 2014). The theorem unveiled that "the definition of rights and obligations is an indispensable prerequisite for market transactions". But the proposition of realizing optimal allocation of resources in traditional transactions ignored the prerequisites. The Coase theorem clearly defines property right and points out that the low enough transaction cost is the prerequisite for market trading. According to Coase, if there are zero transaction costs, the efficient outcome will occur regardless of legal entitlement". The school of thought pioneered by Coase is later known as neoinstitutional economics. After the 1970s, the influence of this school of thought has grown and its theories are regarded as major revision of the neoclassical economics. Property right economics is one of the linchpin of neoinstitutional economics, which is also termed as 'Property Right and Transaction Cost Economics' or even 'Modern Property Right School of the West'. This shows the nucleus position of the property right theory in neoinstitutional economics (Alchian & Demsetz, 1972; Barzel, 1989; Coase, 1960; Demsetz, 1967; Williamson, 1977, 2000).

The gurus of modern property right school air differences of views on the definition of property right. Demasetz's (1967) *"Toward a Theory of Property Rights"* is the earliest classical document about property right economics. He defines the property right as right to benefit or harm oneself or others. Alchian in his *"Property Right: a Classic Note"* gives a brief definition that a property right is exclusive authority to determine how a resource is used (Alchian & Demsetz, 1972). North in his "Structure and Change in Economic History" defines the property right as "in essence, an exclusive right". The entry "common property right" contributed by Cheung to the "New Palgrave Dictionary of Economics" published in 1987 defines it as "competing rules established to settle the conflicts in the scramble for rare resources in human society, which may be laws, regulations, habits or ladder positions" (Cheung, 1983, p. 427).

Despite the differences in the definition of property right, most of the scholars of the modern property right school stress the exclusiveness and traditional nature of property right, holding that constraint of property right usually harm the efficiency of resources allocation, hence the fewer the constraints, the better it is. The tradability of property right is very important. Property right is defined as "a series of rights transferable by owners" (Perman, Ma, & McGilvray, 1996, p. 104). Modern property right advocates have also become aware of the essence of property right as reflecting the behavioral relations between individuals. It is asserted that property rights system can be considered as sets of economic and social relations that define the position of each individual with respect to the utilization of scarce resources (Furubotn & Pejovich, 1972; Furubotn, 2005). Chinese scholar Shaoan Huang (1995) has given a more succinct definition that property right is a right to property, a group of divisible rights to property, including ownership, use right, usufructory right and right to disposal; they are a set of economic right relations of the people (main subject) around or through property (object); perceptually they are relations between man and things, but in essence, they reflect the relations between man and man (Huang, 1995, pp. 55–106). According to Huang's definition, property rights are rights to property.

The complete property rights are the complete rights to a given property, but not a particular right. They are a group of rights or a right system, which is a bundle of property right in property right economics. In most cases, property rights are incomplete or attenuated due to imposed constraints and that is why the "bundle of rights" owned by the right holder is often defective (Schlager & Ostrom, 1993). At present, the most used division is "possession, use, benefit and disposal". Nevertheless, in economics study, the most often used is the simple division that is the right to use, right to benefit and right to transfer (Cheung, 1983).

There used to be a big dispute in China's theoretical circles over the relations between ownership and property rights. After extensive reading of literature and analysis, the view that property right is ownership is feasible. The differentiation of the two comes from the Chinese translation and the habitual use handed down from history. The first is the confusion in translation. Some people translated the English term property rights into Chinese *chan quan*, which literarily means asset rights and some people translated it into *suo you quan*, which literarily means "the right to own". The same term has two versions of translation. Some people believe that they are different because the Chinese term *suo you quan* can be translated into English as ownership. This, in fact, confuses the term *suo you quan* between its broad and narrow senses. "Ownership" in English refers the lawful title to something in its narrow sense, that is, the state of being belonged to or possession and proprietorship. It is a right at the same level as right to possess, right to disposal and the right to use but alienable from them. Property rights, if translated into *suo you quan*, means ownership in the broad sense, including all the powers and functions. Before the Western economics spreads into China, the mainstream economics was Marxist economics. In Marx's works, the team property rights are translated into *suo you quan* and the version has been handed down. That is why property economics is also called ownership economics in China.

What needs special explanation is the relations between de facto property rights (property rights actually exist in realities) and privilege (right to property recognized and protected by law". In discussing property right among the theoretical circles, there is a starting point with hidden dangers, that is, the view that property right is a legal form of economic relations and property right exists only in the sense of law and property rights is privilege or rights in the legal sense. De facto property right is an objective economic right while privilege is a right codified by law. De facto property right is a group of behavioral right of the subject aimed at benefiting from the asset, with both the behavioral interests of the holder of the property right and the objective assets of the property right existing independently or independent of law. Objective property right relationship is the core of social and economic relations, belonging to economic base. When such objective property relationship is acknowledged and protected by law, it has become rights relationship in the legal sense, that is, it has acquired the form of legal rights, which belong to superstructure or ideology. Legal rights must be based on objective property rights, but property rights will not necessarily acquire timely and fully the form of legal rights (Huang, 1995, p. 73).

Neoinstitutional economics further points out that the de facto property rights and de jure rights are not in alignment. Objective property rights will not necessarily acquire the form of legal rights due to lag of making law, and property right definition needs resources, that is, paying the so-called "transaction cost", which is far complicated in de facto property right than in de jure rights. Due to the existence of transaction costs, property rights have never been fully defined as an economic issue and the undefined property rights have been pigeonholed in the public domain and captured by individuals (Barzel, 1989). Property right system in which the law provides that all means of production belongs to the public cannot eliminate de facto private property rights because of the existence of "capture" (Zhou, 2000). In their studies of fishing resources, Schlager and Ostrom (1993) discover that in some circumstances, resource users make common definitions and exercise their rights, which are de facto rights if not recognized by the government. Users of resources with de facto rights act just as those with de jure rights and exercise such rights among themselves. As a factor influencing behavior, de facto rights, if not challenged, is like de jure rights and only when they are challenged, significant differences would appear (Mcginnis, 2000, pp. 118–1221).

The cost of establishing property rights legally, including making and enforcing laws, has also to pay transaction costs. Even if the law on defining property rights is made and the cost of enforcement is inadequate, the de facto property right will still deviate from the provisions of law. Just as Barzel comments, fully defining the right to an asset is so expensive in terms of transaction cost that it would become worthwhile to leave part of the rights to the public domain. Transaction cost economizing is the main reason for the existence of a huge amount of de facto property rights that are not identical with de jure rights. De facto property rights are subject to restrictions by many factors. Law is only one of them. The most important is the objective economic factors. De facto property rights can be independent of de jure rights. The starting point of studying property rights by economics is the objective rights relationship and it does not have the legal sense. The tasks of economics are to reveal the property rights relations that exist objectively and to provide the basis for making policies and laws. The task of this book is to unveil the objective property rights relations around water resources. It is a task of jurists to use law to regulate the property rights relations of water resources. The unveiling of the property rights relations of water resources may provide the foundation for jurists to carry out their work.

This book defines property rights as rights to make decisions, which is a right of option. The right of option about the beneficial use of resources is the criterion for judging whether an entity holds the property rights. This interpretation comes from the definition of property rights that are socially compulsorily enforced rights to choose from among multiple purposes of resources (Alchian & Demsetz, 1972). Built on this view, it is argued that preference is right (Zhou, 2002). Based on this understanding, de facto right to choose (decision) is de facto property right. The right to make decisions is multi-various, such as to select the multi-purposes of resources, method and quantity of utilization, method for getting benefits from the resources chosen and the targets of resources transfer. So, the right to choose is in fact the right alienable from the bundle of property rights. The right to choose is an expression of the will of the property right holders and this can be regarded as the essence of rights. This book also considers the policy decision maker and policymaking entity, which may be individuals or groups, as the de facto entity that exercises the right to choose in the utilization of resources.

1.5.3 Water Rights

Water rights are the rights and obligation relations that objectively exist. So long as there is water governance, development and utilization, there is the problem of water rights. This book defines water rights from both the broad sense and the narrow sense. In the broad sense, water rights refer to the right to make decisions concerning activities associated with water affairs, such as building water projects, prevention and control of floods and waterway shipping. It reflects the rights and obligations of all kinds of policymaking entities in water affairs. In the narrow sense, water rights refer to the property rights of water resources and the power of making decisions concerning the use of water resources, such as allocation and utilization. It reflects the rights and obligations of all kinds of policymaking entities in the use of water resources. As water affairs cover many aspects, the use of water resources is but one aspect. That is why water rights in the narrow sense are contained in the water rights in the broad sense. This book mainly discusses about water rights in its narrow sense. So the term 'water rights' used in the book means the narrow sense of water rights unless otherwise specified.

Chapter 3 of this book goes into details about the implications of the property rights of water resources. Here is a brief introduction. Water rights are rights to property concerning the use of a certain amount of water resources when water resources are scarce. Water resources feature mobility and recyclability. The object of water rights is a certain amount of water. But in essence, it is a bundle of property rights concerning the use of a certain amount of water resources. As the bundle of property rights can be divided, the term 'water right' may refer to a single right in the property right bundle or it may be used in the plural form to refer to the entire bundle of rights. There are many ways of dividing 'water rights bundle' (Schlager & Ostrom, 1993), and this book, according to the characteristics of water resources, holds the view that the bundle of water rights can be divided into right to allocation, right to withdraw and right to use. The three kinds of rights may be relatively divided into two parts: public right, that is, the right to allocation, expressed in political rights; and private benefits, which refer to the right to withdraw and the right to use, expressed in economic benefits. The political rights exercised by the state may be regarded as part of the rights to property, or the overlapping of the administrative right and the right to property. This is determined by the public nature of water resources.

From ancient times to the present, China is a society whose water rights are owned by the public, and the political power exercised by the state concerning water has been stressed over and over again. The private benefits from water is regarded as being contained in the public interests, with the private rights to water benefits long attenuated and water rights in the sense of private property long been ignored. This is the case especially during the period dominated by the planned economy. The country's reform and opening up has boosted the demand for private property rights of water resources. On the one hand, the scope of water scarcity resulting from economic development has been rapidly expanding and competition has grown in the use of water among different regions, departments and groups, hence the stronger attributes as assets of water resources; on the other hand, market economy has created many independent interest groups and a diverse of property right entities and investors, which all claim for the property rights of water resources. In this new social context, the stress of the political power of the state over water is far from meeting the requirements for settling conflicts arising from the use of water, for raising the efficiency of the use of water, for optimizing the allocation of water and for encouraging investment in water development. It inevitably demands greater respect for the water property rights of various interest groups. So the use of the term "water rights" is unavoidable. It also shows that the current heated discussion about water rights is in essence the right to water resource in the sense of an asset. It is, therefore, appropriate to define water rights as property rights in water resources.

As China does not have a law concerning property rights in water resources, it is traditionally regarded as "ambiguousness in water rights". But the fact that there is no law to define property rights in water resources does not necessarily mean there are no property rights in reality. The de facto water rights in the economic sense has already been there and held by policymaking entities at all levels from the state down to the commons. Generally speaking, the government holds the right to macro allocation; groups hold the rights to withdraw and water supply; and users hold the right to use. Government institutions and enterprises, farm households, families and individuals are the end users of water. No matter how the water rights are allocated, they still exercise the water rights. The collective action in the development and utilization of water resources determines that water supply is an organized group behavior. Groups need to obtain the right to withdraw water while allocating the water withdrawal permits within their groups and therefore they need to have the right to allocating water. Water resources are strong in their public nature. The government does not only exercise public management in the general sense but also property management right directly.

This shows that ambiguousness of water rights is only ambiguous in the sense of de jure rights. This may lead to two problems. One is the fairness. If institutional rules are absent, the definition of water rights according to the natural power possessed by various regions or groups will lead to unfairness in the allocation of rights and even to conflicts. The second is efficiency. Rights that are not protected are apt to be challenged, thus they are feeble in stability. It would add the trans-action costs in the turnover of rights. So the efficiency in the allocation of resources is usually low. To solve the two problems, it is necessary to get a final clear defi-nition of rights legally or confirmation, regulation or further definition of the rights relations in the sense of economics. We call this process 'clarifying water rights' or 'water rights definition'. Clarifying water rights is conducive to raising the effi-ciency of the utilization of resources and optimizing the allocation of resources, limiting the extensive growth of demand and easing the pressure on resources and investment and preventing and reconciling conflicts over water use.

In China, the law provides that the state owns all the water resources, that is, all property rights in water resources within the territory of China belong to the state. The state may, according to needs, choose proper methods to exercise the property rights, granting communities and groups the right to withdraw and water users the right to use water. First, the state and its agencies directly develop and utilize water resources and supply it to users by a certain form (usually free supply or compen-satory sale). Second, the state may grant the right to withdraw or use to certain groups or users by administrative means, which is now in force in the country. Third, the state may, through leasing or auctioning, grant the rights to withdraw or use to certain groups. In whichever case, the state remains the administrative manager of water resources (holding the right to macro allocation of resources) so as to ensure the exercise of property rights conforms to public interests. The state exercises ownership of water resources for the purpose of ensuring the maximum public interests and putting the resources to integrated use.[2]

1.5.4   Water Market

The term 'water market' used in the book refers to the water rights trading market, a product of water rights transferring among equal policymaking entities, hence it also called water rights market, a way of realizing the turnover of rights by price mechanism. Theoretically, there are water markets at all policymaking levels, such as trading among individuals, groups and even among governments or countries. The higher the level of the market, the more the turnover of rights involves political procedures and therefore it is political trading in essence. According to the nature of rights transferred, we usually divide water rights trading into short-term and long-term trading. Policymaking entities hold a given right, which constitutes the basic prerequisite for water rights trading and on this basis, there may evolve a variety of

---

[2]Worldwide, water resources are not always owned by the state in the legal sense. In federal states, such as the United States and Canada, water resources are usually defined to be owned by state governments and the cross-state water bodies or watersheds come under the jurisdiction of the federal government. Whatever the legal form, the public nature of water resources determines that the private and public interests must be identical in the exercise of property rights in water resources.

complicated forms of water markets, such as leasing, water bank and even on-line trading.

The application of market in the allocation of water resources and the water market are two different things. Theoretically, once the price mechanism is introduced, market force begins to play its roles to a certain extent. But the application of price mechanism does not necessarily entail water market referred to in this book. In the staged allocation of water resources, the policymaking entities at higher levels may introduce a certain form of market mechanism to allocate rights among policymaking entities at lower levels. For instance, in allocating cross-watershed water rights, the central government may tie investment in projects to the quantity of water to be allocated. In managing water withdrawal permits, management departments may collect water resources fees. The government may even directly put the water rights to auction. All these mean the introduction of market mechanism in one way or another in the allocation of rights at all levels, but it does not necessarily mean there are physical water markets. Water markets are for equal policymaking entities to trade in rights, as there is only market trading in the real sense among the trading of equal entities.

It is necessary to distinguish between water rights market and water supply market. In urban water supply, water is treated as a commodity and users have to pay for the right to use water. This is a market with price as the intermediary, and it occurs among water supply organizations and users, belonging to the allocation among upper and lower levels at the group and user levels, however, there is no trading among equal entities. They are, therefore, not regarded as water rights market. The water rights and the rights regulation among users are mainly controlled by water price and there is no trading among users. In fact, the operation of water supply system is a kind of water allocation at a very low transaction cost, thus saving the cost for clearly defining water rights among users. The study in this book ultimately shows that the expansion of the future markets in the allocation of water resources and the water markets will mainly develop in the water supply market but not in the water rights market. It is, therefore, very important to know the differences between the two. Water market is water resources market. As defined in the preceding paragraphs, water resources include rivers, lakes, ground water, artificial rivers, lakes, reservoirs, ponds, pools and canals. I oppose the division of 'primary water market' and 'commodity water market' as well as the proposition that water rights market is primary water market, because the differences between primary water in the sense of natural run-off and the commodity water at the cost of human labor have become more and more obscure under modern conditions. For instance, a dozen water storage and regulation reservoirs have been built on the trunk of the Yellow River and that has made it possible to realize the water resources control in the whole watershed. So the water allocated along the Yellow River contains huge human and material investment. The key to identify water markets is not what kinds of water traded but who trade them and whether or not it is the trading among equal entities, as well as the nature of the rights traded as de jure rights or de facto rights, short-term rights or long-term rights.

1.5.5 Water Governance Structure vs. Water Rights Structure

The distribution of water rights in the broad sense is called "water governance structure" in this book. Correspondingly, the book defines water rights in its narrow sense as "water rights structure", that is, the structure of property rights in water resources. The water governance structure reflects the rights and obligations relations in the use of water resources. Just as the water rights in its narrow sense is contained in the water rights in the broad sense, water rights structure is in reality also contained in the water governance structure, that is, water governance structure is the macro rights structure while water rights structure is more specific property rights structure. While having its general features of water governance structure, water rights structure also has its own particular features.

Water rights structure is the main target of study in this book. Coase Theorem (Coase, 1960; Gjerdingen, 2014) holds that if there are zero transaction costs, the efficient outcome will occur regardless of legal entitlement, as the society can easily divide all kinds of property into pure private property. Through the rights turnover among equal policymaking entities, the economic agents can bargain to achieve a Pareto optimal allocation of resources. On the contrary, if the transaction costs are very high, the allocation of property rights would become extremely important and property rights structure would have decisive impact on the performance of the allocation of resources. Water resources are exactly the resources that have such high transaction costs. This is the theoretical foundation for me to study water rights centering round water rights structure.

The book has adapted the "institutional hierarchy model" developed in natural resources institutional economics and, on this basis, developed the "water rights hierarchy conceptual model", which reveals the evolution process of China's water rights structure from the ancient times to the present. Of course, the book has not stayed at the description of the water rights structure. It has provided the economics theories for explaining the water rights structure and its changes and analyzing the inherent factors for the changes in the water rights structure during different historical periods. The reader may find that, through a multi-level institutional optional model, it is easy to get a full understanding of the water rights problem. The economics theory of water rights structure may help understand the internal mechanism of the option and change of the water rights system. The empirical study of the historical changes in water rights structure may help deepen the understanding of the changes in water rights system.

1.5.6 Hierarchy

The term 'hierarchy' is coined in the academic circles in the West and it is difficult to know when the word was introduced into China. But it is believed that the understanding of the word comes from Max Weber's theory of bureaucracy, which is an organization best suited to exercising legal authorities. In fact, bureaucracy is not confined to bureaucratic organization. It is the basic framework of many large modern social organizations, such as government, army and enterprises. There are some key features of bureaucratic organizations. Firstly, bureaucracies had a formal and unambiguous hierarchical structure of power and authority. Secondly, bureaucracies had an elaborate, rationally derived and systematic division of labor. Thirdly, bureaucracies were governed by a set of general, formal, explicit, exhaustive and largely stable rules that are impersonally applied in decision making.

The use of the term 'hierarchy' in the book is not based on sociology or political science, but by following the tradition of the literature of institutional economics. In new institutional economics, hierarchy is a form of allocation of a resource vs. the market. The level structure represented by hierarchy often refers to enterprises in organizational economics, but in fact the level structure is not confined to enterprises, but an organizational form corresponding to the market. As an organizational form, the word "hierarchy" used in the literature of new institutional economics has something in common with that used in political science and sociology in terms of hierarchy, administrative orders and compulsory coordination. In this book, the word "hierarchy" sometimes refers to the organizational form of enterprises and sometimes is generalized as a structure similar to the internal power relations of enterprises. Chapter 2 will go into details about the use of the word.

The hierarchy structure holds the key to understanding both China's water governance structure and state governance structure. In the past literature of political science, including those of the "water governance school", have already recognized such special governance structure of China and defined it as 'despotism', 'totalitarianism' or 'bureaucratic system'. By using the term 'hierarchy' to describe China's special governance structure, it is made possible to associate the thinking of political science with the literature of institutional economics. The choice model of governance structure constructed by this book demonstrates the economic logic of the option and operation of the hierarchy structure, that is, the hierarchy structure is the results of economizing the high cooperation costs, in the meanwhile concomitant with fairly high management costs. The maintenance of this structure is preconditioned by effectively lowering the management costs. This is, in fact, an extension of the theory of the firm (Coase, 1937, 1960; Williamson, 1977, 1985, 2000) to the area of political sciences. The book has made full use of this economic logic and borrowed the most recent achievements of new institutional economics to make the "water governance school" achieve new development after years of silence.

1.5.7 Residual Control Rights

Decision-making power, which has been introduced in the preceding paragraphs, has the same meaning as the control power or rights. The rights of residual control are, in reality, the right to residual policymaking power. The term "residual control rights" is often used in the theory of the firm. Advanced by Grossman and Hart (1986), it is one of central ideas of the "incomplete contract" theory and also an important component part of the firm contract theory. According to the understanding of Weiying Zhang (1996), rights of residual control have the following three aspects: (1) the contract nature of enterprises. Enterprises are the combination of a series of contracts and a way of property rights trading among individuals; (2) incompleteness of contract. An enterprise is an incomplete contract. In a world filled with uncertainties, it is virtually impossible to foresee all the possibilities at the time of signing the contract. For an enterprise to survive in an uncertain world, it has to cope with all contingencies and perfect contract is no different from denying the existence of the enterprise; (3) rights of residual control of enterprises concomitant with the incompleteness of contract. Due to the incompleteness of contracts and it is impossible to foresee all the possible situations at the time of signing the contract and the rights and obligations of all parties concerned to the contract, when a real situation appears, there must be people to decide how to stop the loopholes in the contract. That is the original of the rights of residual control. If a contract is complete, there would be no rights of residual control.

Although the concept of residual control rights is advanced in the theory of the firm, it is later applied to the study of hierarchy structure of social organization (Lake, 1996). In reality, the holding of residual control rights is a general method for judging the hierarchical levels of an organization. Usually, the more the residual control rights are concentrated in the upper levels, the higher the degree of hierarchy. In order to facilitate the understanding of this point, we may regard any organization as a unit tied up by "contract", with the policymaking entities at all levels holding the decision-making rights and exercising the power of contract according to the established rules. As a contract is incomplete, there are decision making powers not specified in the contract. If the rights of residual control are concentrated in the hands of the upper level of an organization, the degree of hierarchy is high or concentrated. Otherwise, if the residual control rights are scattered among policymaking entities at the middle and lower levels, it means the degree of hierarchy of the organization is low or decentralized. In this book, residual control right is taken as a theoretical tool for gauging the degree of hierarchy of the governance structure. The essential feature of the hierarchical structure is that the most upper level monopolizes all the residual control rights.

Chapter 2
Hierarchical Structure of Water Governance

This chapter first of all reviews the organizational theory of new institutional economics in order to pave the way in literature and methodology for the study of water governance structure. Then, it proceeds with the basic implications of water governance and finds that the most important output of water governance is to ensure water security. The collective actions for this purpose result in a continuum of governance structure. China's "hierarchy" model is the highest level of hierarchical system in water governance structure, which is unique in the world. It contributed to the unification of the country in the Qin Dynasty more than 2000 years ago and has continued with the unified political system till today.

This chapter constructs a classical choice model for water governance structure that is used to explain the origin of the hierarchy water governance structure. The demand for hierarchy water governance structure is the decisive factor for the formation of the early unified political system in China. This is the view of scholars of the water governance school (Curtis, 2009; Huang, 2002; Needham, 1981; Wittfogel, 1957). By employing the modern economics analytical tools to interpret the relations between water governance and state governance, this chapter enriches and develops the theories of the water governance school.

2.1 Review of Organizational Theory of New Institutional Economics

The path-breaking economic analysis of transaction costs did not occur until 1937, when Ronald Coase (1937) published "*The Nature of the Firm*". Williamson completed the integration of transaction cost economics. From then on, transaction cost economics has not only scored glorious achievements in the theories about the firm but also been applied in much broader areas such as political organization, international relations and the governance of environment and natural resources,

© Springer Nature Singapore Pte Ltd. 2018
Y. Wang, *Assessing Water Rights in China*, Water Resources Development and Management, DOI 10.1007/978-981-10-5083-1_2

thus building up a sizable library with a considerable amount of organizational theories of transaction costs.

2.1.1 Transaction Cost

Transaction cost did not receive enough attention until 1937 when R.H. Coase wrote *"The Nature of the Firm"*, in which he, for the first time, used transaction cost to explain the existence and size of firms. In his article *"The Problem of Social Cost"* (Coase, 1960), he used the concept of zero transaction cost to criticize A. C. Pigou for his logic of state intervention to solve harmful effects of pollution. If transaction costs are zero, Coase argues, then through a process of bargaining an efficient outcome would be achieved without the need for government intervention.

The idea advanced by Coase in his *"The Problem of Social Cost"* was summarized as 'Coase Theorem' (Coase, 1960; Gjerdingen, 2014). According to this theorem, given well-defined property rights and all transaction costs are zero, resources will be used efficiently and identically regardless of who owns them. Property rights and transaction costs are the two sides of a coin, saying that if transaction cost are really zero, the definition of property rights may be ignored (Cheung, 2000, p. 442). Transaction cost is the core concept of new institutional economics. It is exactly the acquisition of positive transaction cost into the framework of neoclassic economics that new institutional economics has gradually revealed the important functions of property rights and economic organizations in the allocation of resources and made politics and institutional structure to become key to understanding economic growth (Thráinn Eggertsson, 2009, 2013).

The accusation that transaction cost lacks accurate definition has never stopped ever since Coase developed the concept. Nearly all prominent new institutional economists have presented their views from different angles. According to Coase, transaction costs are search and information costs, bargaining and decision costs and policing and enforcement costs. Later on, new institutional economists have added more means to transaction costs. Williamson compared transaction costs to friction in physics, which include advanced transaction cost and the costs for signing contracts, defining the rights and obligations of transaction parties and also, after the signing of contracts, the cost for solving problems left over from the contracts per se and for changing articles or withdrawing from contracts. Defining transaction cost from the same perspectives are also Matthews, who holds that the fundamental idea of transactions costs is that they consist of the cost of arranging a contract ex ante and monitoring and enforcing it ex post, as opposed to production costs, which are the costs of executing a contract. Barzel defines transaction costs as costs for acquiring, protection and transfer of rights. Eggertsson (1997) holds similar views as Barzel (1989), asserting that transactions costs are the costs that arise when individuals exchange ownership rights to economic assets and enforce their exclusive rights, namely, exchange cost plus enforcement cost. Transaction costs are argued as costs for the operation of the economic system (Arrow, Sen, & Suzumura, 2011). Similarly, Cheung (2000) views that transaction costs include not only costs for signing contracts and negotiations but also costs for measuring and protecting property rights and costs for obtaining rights and their governing activities, supervision behaviors and organizational costs.

Superficially, differences exist among prominent economists in the definition of transaction costs. John R. Commons (Van de Ven & Lifschitz, 2009), the founder of the contemporary institutional economics pioneered the notion that all human activities may be regarded as transaction corresponding to production activities between man and nature. Commons sums up transaction into three basic types: (1) trade transaction or exchange relations among equal persons; (2) management transaction, which is an exchange relation between superordinates and subordinates; (3) quota transaction, which is a relation between government and individuals. Transaction activity is the basic unit of institution. The operation of institution that is made of numerous transactions and different institutions is nothing but the combination of the three types of transactions according to different proportions (Commons, 1983). Coase takes the first type of transaction as the target of transaction costs and scholars after Commons's explanation of types of transactions (Coase, 1937, 1960). However, the definition that transaction costs are costs incurred in market transactions is relatively narrow, while defining transaction cost as institution cost is broad and comprehensive (Barzel, 1989; Thráinn Eggertsson, 1997, 2009, 2013). All these definitions have inherent interconnections, referring to the costs of exchange among people as contract relations that are all costs of people-to-people exchange (Arrow et al., 2011; Cheung, 2000). Further, transaction costs are analysed in relation to complicated governance structure and social context (Challen, 2000; Krutilla & Krause, 2010; McCann, 2013; Schlager & Ostrom, 1993; Williamson, 2000).

Since social organization is a collective of human relationship, transaction costs are important to understand all social organizations. Organizational cost, agency cost and supervision coast can be regarded as transaction costs. Factors determining transaction costs are complicated. If human activities are regarded as "transaction", then all factors associated with humans are subject to the impact of transaction costs, such as the behaviors of participants, including bounded rationality and opportunism; transaction, including frequency, asset specificity and uncertainty; governance structure, such as market, hierarchy and the combination of the two, legal system, government control and public bureaucracy; institutional environment, including property rights, contracts and culture.

2.1.2 Transaction Cost Economics

Neoclassic economics after Adam Smith regards price mechanism as the most effective in regulating transaction and market as being better than centralized organizations in the allocation of resources. In his 1937 paper, R.H. Coase raises two problems: one is why firms have emerged and the other is what determines the size of a firm (Coase, 1937). He comments that the use of price mechanism needed costs, including the costs for discovering price and negotiations. Transaction costs are the fundamental factor that leads to the existence of firms. Firms employed hierarchical directions to replace voluntary market transaction. The size of a firm is determined by transaction costs. There is a critical point between firms and market, at which the marginal cost of a transaction organized internally is equal to the marginal cost of the transaction concluded on the open market. Coase is the founder of transaction cost economics. His contribution lies in the revelation of the decisive significance of positive transaction costs in organizational choice (Coase, 1960).

What Coase pioneered has been pursued by many new institutional economists. In the 1970s, Williamson picks up Coase core argument and developes it into what is known as 'transaction cost economics'. He holds that, concomitant with the exchanges among individuals with bounded rationality and opportunism in an environment filled with uncertainties is of necessity the transaction costs that often lead to organizational failures. He makes transaction as the basic unit of analysis. He points out that when transaction costs are too high, market will no longer be the most effective governance structure and it is necessary to design a property governance structure (Williamson, 1977). Different transaction costs entail different governance structures. The size of transaction costs requires the study of the specific features of transactions, including frequency of transaction, problem of asset specificity and uncertainties of transaction. Different from Coase, Williamson gives particular stress to transaction costs associated with asset specificity.

Coase regards the power of planning and employer and employee relations as the essence of firms and the activities in the absence of such power and transaction in resources by signing contracts independently as the essence of the market (Putterman & Kroszner, 2000). Coase courts opposition from some new institutional economists for diametrically separating firms and market. Cheung points out that "market transactions involve products or commodities ... 'firm transactions' involve factors of production, and the replacement of a product market by a factor market" (Cheung, 2000, pp. 240–264). Alchian and Demsetz hold that market contracts and firm contracts have de facto continuity (Putterman & Kroszner, 2000). Benjamin Klein also holds that Coase mistakenly distinguishes the transactions within enterprises and among enterprises, adding that the latter is a market contract while the former is a planning one. Economists have become aware that such distinction is so meaningful, suggesting that transaction should be considered within enterprises as market and contract relations (Klein, 1983). However, economists have not reached a common understanding and the disputes will continue over whether or not a firm is an entity under administrative directions without being influenced by the market force or merely an entity linked up by a series of contracts signed among individuals.

Miller (1992) argues that hierarchy and market are different in nature. The coordination within firms is mainly realized by hierarchical instructions, and the transition from market to hierarchy has enabled individual behaviors to undergo tremendous changes. Hierarchy realizes coercive coordination by administrative orders. The significance of the criticisms against Coase lies in the revelation of the contract nature of the firm, that is, the firm involves the series of long-term contract relations among those who have input factors and the firm tends to replace factor market with product market. The benefit of regarding the firm as a contract association of resources holders lies in the diversity of the forms of associations of resources owners, which can be explained from the cost economizing feature of different contract arrangements. Nonetheless, it is also necessary to see at the same time that the price signals on the factor market has little roles to play and it is often the case that hierarchy, i.e. power relations, has replaced market exchange (Miller, 1992).

2.1.3   Market vs. Hierarchy

Figure 2.1 shows the differences between the two kinds of governance structures of market and hierarchy. In the figure, every arrow represents a transaction and this is the basic unit of analysis. The relations of any two individuals form a contract and one contract may include a number of transactions. A contract may be formal or definite and may be non-formal or hidden. In this framework, market and hierarchy are merely the connections of contracts. The only difference lies in the contract organizational structure. Hierarchy and market are not necessarily simple two divisions. Most scholars are inclined to hold that the firm and the market are incrementally changing from one form to another. In order to characterize the incremental changes of the two structures, we view the tendency of organization away from hierarchy but closer to market 'leveling off'; while the tendency of organization closer to hierarchy but away from the market 'hierarchy'.

Different transactions are conducted in different organizational forms in that the society always favors organizational structures that are most conducive to transaction cost economizing as explained by transaction cost economics. The transaction

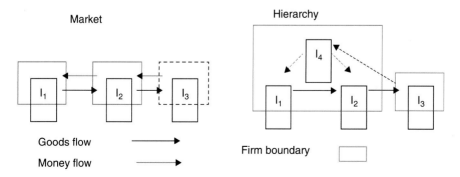

Fig. 2.1 Comparison of market and hierarchy structures

cost economizing by replacing hierarchy with market comes from many aspects: (1) the cost for searching market information; (2) cost for negotiation and drawing up contracts; (3) cost for enforcing the contracts and settling conflicts; (4) cost associated with market incompleteness, which might come from externalities, natural monopoly, information asymmetry and asset specificity; (5) scale economy and scope economy possibly brought about by the firm. The existence of hierarchical firms is the result of market efficiency failure, which may be resulted from the above factors (Beckmann & Padmanabhan, 2009).

When market fails, hierarchy may sometimes improve the low efficiency of market, but not always so. In reality, there is also hierarchical failure, associated with low efficiency. This is because, while reducing the transaction costs, there might be new costs arising with hierarchy, that is, the costs for forming and maintaining associations of producers, which is usually termed 'agency cost' (Thráinn Eggertsson, 1997). Agency cost come possibly from the following: (1) cost of supervision by principal over agents; (2) efficiency loss associated with "moral risks" of agents due to incompleteness of supervision; (3) costs of coordination between different agents; (4) efficiency loss due to incomplete information associated with information filtering.

The firm or the market, each organizational form has transaction costs and the size of the costs is determined by the degree of mutual replacement. As the firm expands, the agency cost within the firm may increase and when the internal marginal gains (cost for negotiation and price discovery is reduced and the problem of asset specificity is eased or other factors) are equal to the marginal cost of integrated management, and the firm has reached the optimal scale. More generally, economic organizations are designed, through identifying transaction features, to be a governance structure that seeks transaction cost economizing.

2.1.4 Expanded Application of Transaction Cost Methodology

While transaction cost theory has achieved tremendous successes in explaining economic organizations, its methodology has been extensively applied in the study of non-economic organizations. In fact, in his paper "The Problem of Social Cost" in 1960, Coase states that transaction cost methodology can be used to explain all kinds of social arrangements, including market, the firm, government and their evolution (Coase, 1960). New institutional economics also puts forward 'General Coase Theorem' that when transaction costs in political and economic areas are zero, the economic development of a country is free from the influence of the government; however, if there are positive transaction costs, the institutional structure that allocates and formulates rules by the political power of a country will become a key factor in determining the economic development. The General

Coase Theorem illuminates that transaction cost economics has explanatory power over non-economic organizations (Gjerdingen, 2014).

In terms of states, a theory of defining how states implement property rights, which considers transaction costs as a variable of state behavior (North, 1981, 1990). Moe (1984) publishes his article "Organizational New Economics", suggesting that since new institutional economics has made revolutionary contributions to the understanding of hierarchical structure in the private areas, it should also be conducive to understanding the hierarchical system of public departments. North (1990) uses the analytical framework of transaction cost economics to analyze political problems. After more than 10 years of literature accumulation, what is termed 'transaction cost political science' has taken shape (Ma, 2003). According to Ma Jun, what the transaction cost political science concerns about is how the transaction costs in political deals influence the design of political system. As it is built on the Western soil, what this branch of science answers are all problems about Western democracy, such as why rational voters choose to vote? Can political principal effectively control bureaucracy? How does parliament take authorization decisions? How to overcome political opportunism effectively? Why some public service can only be generated within public bureaucratic organizations? These subjects of study are far from the Chinese social structure and political activities and for such discussions this book will not go into details.

Transaction cost methodology has also been applied in the studies of international relations and politics, such as why countries adopt different forms in cooperation with different countries. Lake (1996) uses transaction cost theories to explain why some countries have resorted to relatively loose cooperation in the production of state security while others prefer hierarchical cooperation. Lake points out that in the process of the production of state security, a state must exchange with other countries and therefore opt for a certain form to govern the exchange with other countries, which ranges from non-governmental and loose alliance to protectorate and non-official empire and to highly hierarchical empire, which form a continuum. In his model, Lake uses governance costs and expected costs of opportunism to explain the options for country-to-country relations. The governance costs rise with the rise in the levels of hierarchy while expected costs of opportunism rise with the lowering of the levels of hierarchy, with the two going together to decide on the optimal relations in the exchanges of countries (Lake, 1996).

Weber (1997) constructs a transaction cost model for inter-country cooperation by using the two variables of external threat and transaction costs. The bigger the external threat, the closer the alliance countries tend to develop; the higher the transaction costs, the more inclined the countries tend to opt for non-cooperation or non-official alliance. Weber's model is in fact much similar to Lake's. Its variable of external threat is amounting to the expected costs of opportunism of the latter and transaction costs are amounting to governance costs of the latter. These achievements are of great significance for this book to draw on.

Since the 1990s, transaction cost methodology has widely been applied in natural resources and environmental economics. Before that, this branch of sciences mainly adopted the neoclassical economics method. With the introduction of transaction costs, a new branch of sciences 'institutional environmental and resource economics' (IERE) has taken shape. The founding work of IERE is Bromley's property rights system and Ostrom's studies of public pool of resources (Bromley, 1995; Bromley & Segerson, 1992; Ostrom, 1990). On this basis, Ray Challen (2000) develops a conceptual framework for transaction costs and institution options, which provided a method for the comparative study of different institutional structures by using cost-effectiveness framework, which will be introduced and quoted in later chapters of this book.

2.1.5 *Effectiveness of Transaction Cost Organizational Theories*

The extensive application of transaction cost methodology reflects in a way the expansionist nature of "economics imperialism". Why has the transaction cost methodology been so widely applied in all economic, political and social organizations? Williamson (1985) has given the answer when he asserts that the basic unit of analysis is transaction and contract and any relations concerning contract may be approached from the angle of transaction cost economics. Economic activities, political activities and the relations among regions or countries may, in fact, be regarded as a kind of transaction among individuals or groups, which can be examined from the angle of contract. Statesmen and voters may be regarded as a contract relationship and statesmen and bureaucratic organizations may also be regarded as a contract relationship. International relations may be regarded as contract relations among different countries; the relations between the central and local governments can also be regarded as being bounded by contract. Some of the contract relations are clear but most are hidden, thus bringing about huge transaction costs in drawing up and performing the contracts and affecting the design of organizational form and institution. So, transaction cost theories are useful in analyzing all kinds of social organizations (Ma, 2003).

The core of transaction cost organizational theories is how to realize transaction cost economizing governance structure. This depends on how people are organized and on what scope it can produce the maximum cost-effectiveness. However, efficient organizations will not form automatically. The efforts to reduce transaction costs depend on external pressure. The effectiveness of economic organizations depends on the competitiveness of the market; and the competitive environment may make efficient economic organizations to replace inefficient economic organizations. North's statement on behavior theories have revealed that a country must maximize not only rents of the ruler but also social output by reducing transaction costs. The two objectives are conflicting with each other. It is exactly the abuse of the first objective that has caused the non-competitive environment and led to the existence of a large number of inefficient economic organizations, hence the decline of countries in international competition (North, 1981). Other social organizations follow the same logic. The effectiveness of a political organization is determined by how big a pressure for reform the political ruler has to bear and what efficient organizations to develop. The efforts by a country to develop effective international relations are determined by how grave the challenges the country faces in state security.

Since transaction costs economizing does not come automatically and it is associated with external circumstances, an organization that can minimize transaction costs is the imagination in the ideal sense. But how to look at the effectiveness of all organizations in realities? Any organization has its case for existence and its aim is to undertake one or multiple tasks and that will force the option for a form that can economize costs. An organization may not perform well due to insufficient external pressure. If it cannot resolve the established problems for long, it would lose its ground for existence. If all the constraints, including external pressure, existing technology, information cost and future uncertainties, are taken into consideration, people would always opt for effective organizations from the static approach as among all optional structures under constraint conditions, the transaction costs of the existing organization have been minimized. What is thought to be an inefficient organization has its roots in that all sorts of constraints and inefficiency. With the changes in constraint conditions, an organization has to respond to external changes and such response determines the organizational changes—to be replaced by a new organization or to die out. The continued existence of inefficient organizations in history may be explained by the 'interest group model' developed by Mancur Olson (Olson, 1971, 1982, 1996). Social organizations are in essence 'institutional arrangements', encompassing a series of rules and conventions, subordinating procedures and moral and ethical behavioral standards (Olson, 1990). The combination of institutional arrangements forms an 'institutional framework', which resonates with North's (1981) explanation of 'structure'. In this book, water governance structure is the institutional framework for water governance, and transaction cost theories play important roles in explaining water governance structure.

2.2 Water Governance in the History and Outcome

2.2.1 Differences of Water Governance in Ancient and Present-Day China

China is a big country in terms of water governance and it has a long history in this regard. The history since legendary Da Yu is also history of the Chinese nation in the fight against drought and floods. Water control in China is large in scale and is of special significance in the continuity and development of the Chinese civilization.

In ancient China, flood prevention, irrigation and navigation were the three main areas of water governance. The first and foremost is the fight against drought and floods. The country is known as "suffering from starvation in every 3 years, from a decline in every 6 years and from a crop failure in every 12 years". It is ranked first in the world in terms of the frequency and intensity of natural disasters, especially floods and droughts. From 180B.C. to AD1949, of all the natural disasters China sustained, more than 90% were droughts, floods, earthquake and tidal waves and the tolls taken by droughts and floods made up 51% of all tolls of natural disasters, averaging 14,210 and 1863, respectively, a year. The economic losses were inestimable (Wang & Tian, 2010). The country frequently built large scaled water projects to resist droughts and floods. In ancient time, the country committed a great deal of human, materials and financial resources to the building, maintenance and protection of dykes along major rivers, especially along the lower reaches of the Yellow River. China's farming was highly dependent on irrigation and the number of irrigation projects was far more than in any Western countries. The number of irrigation project before the Tang Dynasty averaged 10 to 16 for every 100 years. In the more than 1300 years after the Tang Dynasty, the number of such projects rose sharply, with the areas under irrigation in 1400 and 1820 accounting for 30% of the total arable land while the proportion in India in 1850 was only 3.5%. There is no country in the world other than China that has committed such human and financial resources to the building of water projects. Waterway shipping also occupied a significant position in China's water control history. The most important shipping project that linked the south with the north is the Grand Canal. Work on the canal started in the Sui and Tang periods. It links Qiantang River with the Yangtze River, Huaihe River, Yellow River and Haihe River. It is a major artery that links south and north in the Chinese history. It is a lifeline that has made north China able to maintain its position as political and cultural centers. The Grand Canal served as a waterway hub for nearly 1000 years and did not phase out until the end of the Qing Dynasty, when land and marine shipping began to pick up (Shen, 2014). After the founding of the P.R. China, the fight against droughts and floods remained the main area of water control. Statistics show that droughts and floods assumed an upward trend (see Table 2.1) from 1949 to 2000 in terms of areas stricken, disaster areas and disaster rate. Droughts and floods have caused huge losses. The areas of farmland stricken by droughts averaged 20 million hectares a year, causing grain reduction by tens of billions of kilograms, about 50% of the total caused by meteorological disasters. The seven major river basins suffered a flood in every 3 years on average and the areas affected every year reach 7,333,333.3 ha and the grain output reduction accounted for 27.6% of the total output reduced. Economic losses run up to tens or dozens of billions of yuan (MEPPRC, 2016). The 1991 Huaihe River floods and the 1998 three rivers floods caused heavy economic and social losses. In order to fight against floods, the country, during the planned economy period, mobilized mass movements to build water control projects and tamed major rivers at a very low cost. During the period, the country built more than 80,000 reservoirs

Table 2.1 Areas flooded, areas stricken by drought and disastrous areas in years from 1950 to 2011

Year	Flood			Drought		
	Areas stricken (ha/year)	Disaster area (ha/year)	Disaster rate (%)	Areas stricken (ha/year)	Disaster area (ha/year)	Disaster rate (%)
1950–1959	7,891,300	4,962,500	57.53	13,223,800	4,166,300	34.11
1960–1966	9,420,000	5,854,300	57.74	21,647,100	10,025,700	45.80
1970–1979	5,357,000	2,243,000	39.64	21,641,000	7,500,000	28.02
1980–1989	10,425,000	5,529,000	52.71	24,638,000	11,761,000	47.56
1990–2000	14,593,600	9,230,000	63.2	26,322,700	13,318,200	50.60
2001–2011	112,727,000	59,439,000	52.7	239,812,000	133,464,000	55.70

Sources: Calculated based on data in "China Statistical Yearbook"

and more than 200,000 km of dykes, thus ensuring the demand for water by the rapid economic and social development. Unprecedented achievements have been made in flood control of the Yellow River. Irrigation developed at a pace never seen in history, with the areas brought under irrigation from 199.60 million hectares in 1952 to 538.51 million hectares in 2000. The total irrigated areas were raised from 18.5% to 51.8% (1995) while it was 29.5% in India, 11.4% in the United States and 4% in Russia during the same period (WB, 2016). "A good ruler controls floods and drought first"—this saying by Guan Zong of the Spring and Autumn period still stands today (Gu, 1997).

Prevention and control of water pollution is something new in water governance since the country introduced reform and opening up policies. Since the founding of New China, thanks to economic development and the application of modern water control technologies, water governance has been diversified. Apart from anti-flood and irrigation projects, it also includes hydropower, control of waterlogging and salination, water and soil conservation, urban water supply and drinking water for man and animal. With the increase in population and urbanization, water shortage has become acute. Urban water supply has become more and more important. Since the country introduced reform and opening up policies, China has become a country in the world that discharges the biggest amount of sewage water (Jin, Zhang, & Tian, 2014; Yang, Flower, & Thompson, 2013). Water body pollution has become more and more serious, thus intensifying water shortages and posing a direct threat to the health of the people. Water pollution control has thus become a new area of water governance, which has equal importance as controlling floods and fighting against droughts. Up to the beginning of the twenty-first century, the ecological restoration in some river basins has been put on the agenda. By now the ecology-associated water resources regulation projects have been set up in the Tarim, Heihe and Yellow river basins. Water ecology restoration project has also kicked off in the Haihe River basin. Although the water governance in contemporary China has been greatly enriched, with increasingly wider coverage, in general, the prevention and control of floods, fight against droughts and water pollution have remained the three most important tasks for China in the first half of the twenty-first century.

Compared with ancient China, the present-day water governance has acquired an entirely new meaning (Liu et al., 2013).

2.2.2 Outcome of Water Governance

Water is the controlling factor, inseparable from the whole water ecological system. Other factors in water and ecological environment together go to form the natural ecological system, which provides humankind with all service assets, not only food, medicine and other daily use materials but also life support system on which human survival depends. Water on earth is in a cyclic state, including rainfall, evaporation, run-off and evaporation cycle. The land water cycle system directly acts upon humans. As the land water takes watershed as the unit, land water cycle may also be regarded as watershed water cycle system, which is an organic component part of the natural ecological system, with the point where watershed water cycle system and natural ecological system together comprises a natural water ecological system.

Natural water system demonstrated by Fig. 2.2, as part of the ecological system, also provides mankind with a diverse of goods that produce economic products, such as water volume and water energy source; environmental service as water environmental capacity; and ecological service, biological habitat and comfort service. As an asset, natural water system, like ecological system, has the biggest value of providing social and economic system with support service. Such asset has been meticulously utilized and protected. When man and nature exist in harmony, the natural water system provides humankind with all kinds of welfare and support for the healthy development of the economic and social system. If the asset is over-used, as much as that it has gone beyond what the water resources of the natural water system and water environment can take, man and nature would come into conflict, thus hampering the sustained development of the economy, threatening the normal social order and even leading to instability.

The topmost task of water governance is to ensure social and economic development, with the fundamental purpose of keeping balance between socio-economic system and the natural water system in order to realize harmonious development. During different social and economic stages and historical periods, the principal contradictions between social-economic system and the natural water system assume different forms, which determine the requirements for water governance. Generally speaking, water governance has to ensure the requirements of social and economic development at the following levels. The first level is the security of drinking water, that is, to supply clean drinking water to the people. This is the most essential requirement by the society for water. The second level is the security against floods, that is, to protect the lives and property of the people. The third level is food security, that is, to ensure food grain supply, which is especially important to China. The fourth level is water security for the economic and social development

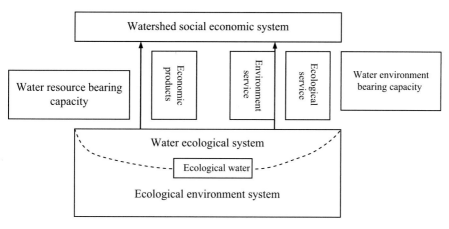

Fig. 2.2 Water ecological system and social-economic system

as economic and social development raise higher demand for water volume and water quality under the condition of satisfying survival. The fifth level is ecological environment security. After the requirements at the above five levels are satisfied, the important task is to improve ecological environment (Yahua Wang, 2013b).

Water security occupies a very prominent position in China. China has carried out long and large-scaled water control in which the natural geographical conditions have made the Chinese society in a state of extremely insecurity. Ancient China made water control as a matter of major importance in maintaining national stability, because floods and droughts were the biggest threat to human survival and the strongest destruction to the society. There were many incidences of social upheaval due to floods and droughts that accelerated the collapse of dynasties.[1] In contemporary China, flood remains a sting in the heart of the Chinese people and water shortages and water pollution have become important factors holding up economic development. That determines that the central task in water governance is to ensure security against floods, security in water supply and water environment and support for sustainable economic and social development. The book makes water security as the objective of water governance structure and the basic point of departure in analysis.

---

[1]For instance, the 16-year-long drought in 1628–1644 caused dramatic drop of crop fields and people to die of famine in 13 provinces and cities including Shaanxi, Shanxi, Shandong, Henan and Jiangsu. The severe drought accelerated the collapse of the Ming Dynasty.

2.3   Water Governance Structure

2.3.1   Importance of Water Governance Structure

Threat by water is a challenge a society has to face up to. In order to maintain survival and development, a society has to take up the challenge. To cope with challenges, the establishment and improvement of institution is fundamental. It is both the result of successful coping with challenges and the prerequisite for meeting new challenges. In order to ensure water security, it is necessary to make long-lasting institutional arrangements and shape up a governance structure good enough to cope with challenges. Governance structure is of vital importance in ensuring water security. Efficient economic organizations hold the key to economic growth of a country, while inefficient economic organizations are the cause of economic stagnation and recession (Song, 2016). Careful examination of the history of China's water governance could easily arrive at similar conclusions as those done by North, emphasizing that governance structure and economic organization are equally important. An efficient water governance structure, though not necessarily ensuring good water governance, holds the key to good water governance while inefficient water governance structure is, of necessity, unable to ensure good water governance and on the contrary even aggravates water disasters or even leads to conflicts or war. The following is a simplified model that can explain the relations between water governance structure and its governance performance.

Water governance performance can be measured by the level of water security ensured. The model assumes that under a given natural geographical condition, the performance is determined by governance ability, which is expressed in the function of three variables: economic, technology and institution. During a given period, the economic strength of a country determines the amount of input into water governance; technical level determines the actual efficiency of economic input; and economic and technical conditions together determine the possible maximum output level in water security, which is expressed in production possibility curve in the model. Figure 2.3 shows the coordinate may be seen as the specific goods produced from water governance such as X is anti-flood security level; Y is food grain security level. The production possibility curve is the combination of the two goods, which reflects the maximum governance ability under the constraint of technical and economic conditions during a given historical period.

It is only a theoretical possibility for economic and technical conditions to determine the maximum governance ability. The actual governance ability is always lower than the possible maximum value, which is located within the production possibility curve in the figure. The model assumes that the actual governance ability is determined by a given institutional framework (governance structure). A good governance structure is useful in mobilizing to the maximum the social resources so that the actual governance ability gets closer to the possible maximum value. Conversely, inefficient governance structure is unable to mobilize

Fig. 2.3 Water governance
structure and governance
ability under constraint
conditions economically
and technically

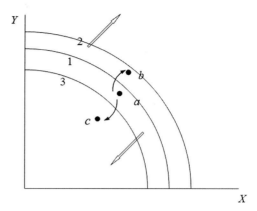

social resources effectively and on the contrary artificially reduces the actual governance ability to make it far away from the possible maximum value. Statically, the model shows that economic and technical conditions during a given period of time and governance structure together determine the actual governance ability and correspondingly the water governance performance in the period, that is, actual output level of water security.

Dynamically, the model shows that the economic and technical conditions are constantly changing. If the economy grows sustainably, the input into water governance would increase and the production possibility curve would move outward (1→2). If the economy is in long-term recession, the input into water governance would be reduced and the production possibility curve would move inward (1→3). Technical progress can improve the efficiency of input in water governance and the production possibility curve may also move outward. The outward movement of the production possibility curve means the opportunities for increasing actual governance ability. If the governance structure makes effective response, the actual governance ability would rise correspondingly(a→b). If the governance structure is unable to be adjusted timely, the actual governance ability would not necessarily grow with the outward movement of the production possibility curve. The inward movement of the production possibility curve may make the original governance structure unable to continue and thus the actual governance ability drops correspondingly (a→c). That is to say, efficient water governance structure may, by mobilizing social resources to the maximum, elevate the actual water governance ability closer to the possible maximum ability while inefficient water governance structure would artificially lower the governance ability, thus dragging the actual governance ability far away from the maximum governance ability. That means that economic and technical conditions can only determine the governance ability, which could not turn into actual governance ability automatically. Only when under a given governance structure, can it turn into actual governance ability. This is the important manifestation of institution.

2.3.2 Water Governance Continuum

Water is a unity composed of basin or hydro-geological unit and that determines that water governance requires overall arrangements of a whole basin. Water governance involves the interests of decision-makers of all regions in the basin. In order to prevent and mediate trans-boundary disputes and reduce negative externalities resulting from cross-boundary damages and expand the positive externalities of cross-boundary cooperation so as to produce economy on a larger scale, it requires all policymaking entities to carry out collective actions. There are two reasons for collective action of different regions within the basin. One is to reduce negative externalities and prevent and mediate disputes resulting from cross-boundary conflicts. There are usually four kinds of disputes over water among different regions: (1) disputes arising from the building of water projects by one region that would affect the rights and interests of neighboring regions; (2) disputes arising from the allocation of water quantity or other rights and interests in the use of water; (3) disputes arising from flood-prevention, water drainage and river course control; (4) disputes arising from water pollution. The second disputes can expand the positive externalities and produce economies on a larger scale. Usually, integrated development, utilization and governance of water on the basin basis according to unified arrangements and planning may bring about greater scale economy. If different regions do not act collectively, the supply of the behaviors with positive externalities would be reduced. This has made necessary the collective action in building reservoirs along the middle and upper reaches, river course control in the lower reaches and in building anti-flood dykes. So long as there are externalities, positive or negative, all demand for collective action exist. The model of collective action of interest groups of different regions in a basin may largely be summarized as the following forms in Fig. 2.4.

The first is anarchy. Every region has its own independent action, without interfering in each other. This model is applicable to rivers that are not so harmful and there are no externalities in the use of water, especially international rivers. In ancient times when productivity was underdeveloped, the governance of many rivers was regional and there was not the necessity for collective action. The water governance of rivers was the affairs of each and every region. With the rise in the level of the development and utilization of water resources, the potential contradictions among different regions in a river basin increases and the anarchical governance model was prone to induce conflicts and even wars and there arose the necessity of united action in the governance of rivers. In the twentieth century, 145 agreements were signed on the governance of international rivers (Wolf, 1998). This shows that the anarchical model tended to be reduced in a modern society.

Fig. 2.4 Continuum of water governance structure

The second is agreement. Different regions in a river basin conclude agreements in the common interests of river governance and carry out collective action according to the agreements. This model has been extensively applied in the governance of international rivers. For instance, Lake Ontario is surrounded by eight states of the USA and two provinces of Canada. They reached an agreement in 1982 that there should be no change of water course without the unanimous consent of the ten local governments. Another instance is the water resources of the Nile River. Egypt and the Sudan signed a water use agreement in 1959, which provided a framework for the allocation of the water resources. Even within a country, this model is also mostly adopted in the cross-regional river basins, especially among federal states. In the United States, interstate agreement is one of the three major ways of easing conflicts in the use of water (the other two are to bring cases to the highest court and the establishment of interstate organizations). At present, there are many single purpose interstate water agreements, such as agreement on the alloca-tion of water resources, agreement on controlling water pollution and agreement on setting up related public organizations. There are also some multiple-objective interstate agreements. The American Colorado River flows across seven states and the allocation of the water resources is mediated according to a series of agreements signed in the recent century (Kauffman, 2015; Schmidt & Shrubsole, 2013).

The third is consultation. In order to better oversee the implementation of river governance agreements or more flexibly resolve transboundary problems, unofficial consultation organizations may be set up to coordinate the action of different regions. This method has been extensively adopted in the governance of rivers, international or domestic. The main functions of the consultation organization are coordination, study and promotion. The policy decision taking must usually get the unanimity of all members. There is almost no coercive power. In the instance of the Nile, Egypt and the Sudan later established a joint committee on the basis of the agreements they signed to solve the water problems of the Nile by consultation. The water pollution control of River Danube in Europe is undertaken by a joint organization formed by a number of countries. Canada has also adopted this model in the management of part of its rivers. Taking Fraser River basin as an example, before the formation of a whole basin management organization, the upper and lower reaches of the river were managed separately. The Fraser Basin Council is a pure non-governmental consulting organization, governed by the unanimity of all regions and communities, without coercive powers. But in the United States, it is only in recent years that formation of interstate organizations has become the trend of the day. For instance, the Mississippi River has set up two non-official organizations to coordinate the interstate water affairs. The Great Lakes area has set up a Great Lakes Commission to coordinate the water issues in the lake area (Kauffman, 2015).

The fourth is coordination. When it is difficult to reach unanimity, it is necessary to introduce a coordination mechanism that is coercive in nature. This model often involves the establishment of official organizations that have varying degrees of coercive powers. This model is often adopted in federal states. The coercive power grows from weak to strong as in the following three cases. In India, irrigation and flood prevention are the affairs of states while the central government coordinates interstate relations by setting up a national water resources committee, the central water committee and the Ministry of Water Resources, which work together to realize the coordination among different states. In the United States, the Delaware River basin and the Susquehanna River basin have set up official river basin councils with a status between federal and state, which have coercive powers over their respective river basins. There are also several other river basins in the United States that have set up basin management organizations that have coercive powers with the approval of the Congress. The New York-New Jersey—Connecticut Interstate Environmental Commission and the Ohio Valley Water Sanitation Committee have the management authorities in water quality.[2] The United Kingdom have set up water management bureaus with independent powers for major rivers to undertake the governance and water resources management of river basins, without being subject to the intervention by local authorities. The biggest among them is the Times River Management Bureau.

The fifth is hierarchy. If the coordination model is not enough to settle the problem of collective action among different regions, then, there is a more powerful centralized model. That is the government or quasi-governmental authorities to directly intervene in the inter-regional water affairs. This book calls this water governance structure with the highest centralization of power "hierarchy" model. Centralized states often resort to this model in river governance, such as the UK, France and Japan. However, due to constant delegation of power over the last century, their water governance often has the characteristics of other models of separation of power. China is one of a few centralized mono-system states and has adopted the pure "hierarchy" model in its river governance. The central government plays the leading role in inter-regional water affairs. Federal states may also adopt the "hierarchy" model in governing some rivers. The best known is the Tennessee Valley Authority, which was established in 1933 with the approval of the Congress. It is like a federal organization, with an extensive coercive power not only in water resources management but also in such regional development projects as power generation, waterway shipping and land resources development.[3]

---

[2]There are seven interstate agreements in the United States approved by the Congress and based on these agreements interstate water management organizations have been set up. They have different degrees of management authorities. The strongest is the Delaware River Basin Council (DRBC, http://www.state.nj.us/drbc/drbc.htm) and the Susquehanna River Basin Council (http://www.srbc.net/),followed by the New York—New Jersey-Connecticut Interstate Environmental Commission (http://www.iec-nynjct.org/) and the Ohio River Valley Water Sanitation Commission (http://www.orsanco.org/). The other three are Potomac River Basin Commission (http://www.potomacriver.org/), the Great Lakes Commission, (http://www.glc.org/), and the New England Interstate Water Pollution Control Commission (http://www.neiwpcc.org/),which have limited coercive powers.

[3]Rivers subject to the management by the 'hierarchy' model can also have basin organizations to coordinate the interests among different regions of the basins. The basin organizations are mainly acting on behalf of local interests. But in the coordination model, the basin councils or other organizations are something in between.

What is discussed above are the basic forms of five kinds of collective action, but they are not diversely different in practice. They are all in a continuum, developing from anarchy to hierarchy, with one level being higher than the other, which is measured by the concept of "residual control power". In the anarchy model, decision-making entities have nothing to do with one another. They have the entire control power over their own decisions. In the agreement model, decision-making entities have lost part of the control power due to agreements, but the residual control power is still held by each group. In the consultation model, a consultation organization settles common affairs through consultation, which is not necessarily based on 'unanimity'. Even there is 'unanimity', some decision-making entities are possibly forced to do so. In fact, they have lost part of their residual control power. In the coordination model, a coordination organization holds a fairly big independent decision making power to bring together the collective actions of the decision-making entities at lower levels, which have lost more residual control power. In the hierarchy model, all the residual control power is held by the decision-making entities at the superordinate level, with those at lower levels losing all their residual control power.[4]

The water governance structure in China is of the hierarchy model in that the residual control power is held in the hands of the government. As China is a mono-system country, the de facto residual control power is held in the hands of the central government. But in practice, management at different levels is necessary. Local governments have also been granted a certain measure of residual control power. In the hierarchy governance structure, the central government is the ultimate principal while local governments are but the agents of the central government. As far as the basic form is concerned, China's hierarchy water governance structure has undergone no fundamental changes over the past more than 2000 years since the Qin Dynasty, with all water related activities, such as waterway shipping, flood prevention and the building and maintenance of major water projects, put under the direct control of the central government. All the canals and waterways and dykes and dams, irrespective who have built them, have always been subject to the management by the government, except for small irrigation projects without large water allocation systems are left to private management. The fact that China has been able to effectively control water and maintain the continuity of the centuries-old civilization is inseparable from the hierarchy governance structure.

---

[4]This is only a single tier analysis instead of multi-tier analysis. The five typical water governance structures mentioned in the book, arising from the level of cross-boundary regions, are applicable to collective action at every level of the hierarchy structure. In the tiered structure, the policy decision making entities at every level may have different models of collective action. For instance, for a cross-country river, the upper-most level is the agreement of national policy decision making entities while at the intermediate level, provincial decision making entities within a state adopt the coordination model. At lower levels within a province is still the hierarchy model. The bottom level adopts the consultation model again.

2.4 Choice Model of Governance Structure

The literature cited above shows that the option for governance structure follows the general logic of transaction cost organizational theories, that is, under circumstance constraints including objective constraints, organizations seek a structure that can minimize transaction costs, i.e. governance costs. Given in the following is a governance structure choice model, which illustrates the inherent logic in the option for water governance structure. The model is also applicable in explaining the option by a state for governance structure.

We define governance structure as variable e, with the increase in e, the governance moves toward the higher level in the continuum. It is, therefore, necessary to define transaction costs that move in the opposite direction. In the governance structure continuum, with the rise in level, the costs also increase. For instance, the agency cost, also known as management costs (C_m). With the level descends, the costs increase, such as negotiation costs, contract performance costs, which are called collectively cooperation costs (C_n). The total governance costs (TC) is the sum total of cooperation and management costs, which are all water governance structure e's function. The definition is as follows:

$$C_m = L(e_i); C_n = I(e_i)$$

$$\text{Of which: } e > 0, \dot{L}(e) > 0, \dot{I}(e) < 0, \ddot{L}(e) > 0, \ddot{I}(e) > 0,$$

$$TC = C_m + C_n = L(e_i) + I(e_i) \tag{2.1}$$

$$\text{The condition of the first order of Min}(TC) \text{ is: } \dot{L} + \dot{I} = 0. \tag{2.2}$$

Define $\dot{L} = MC_m, -\dot{I} = MC_n$; MC_m is the monotonic increment function of water governance structure e; MC_n is the monotonic decrement function of water governance structure e, implicating respectively the marginal management cost curve and marginal cooperation cost curve (absolute value taken). As is shown in Fig. 2.5, the two marginal cost curves cross at e^*, the minimum value point of the total governance costs and also the equilibrium point of the governance structure.

Fig. 2.5 Optimal equilibrium model of governance structure

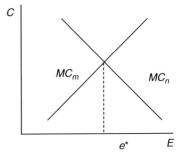

2.4.1   Cooperation Costs

The less tiers of the governance structure hierarchy increase cooperation costs, which include the costs for collecting information about cooperation partners, the cost of reaching agreements and the cost of enforcing the performance of the agreements. Cooperation costs are originated in the opportunism of participants in collective action. For instance, some participants lack the sincerity in cooperation or ask for exorbitant prices in negotiations, resort to deceits or breach of contracts by taking advantage of information asymmetry, or adopt actions harmful to other participants under uncertainties. Such costs in the firm theories are costs of utilizing market mechanism. Here they refer to costs by taking advantage of political negotiations (Thráinn Eggertsson, 2014).

In the model, the cooperation costs are the function of the governance structure, which increases with the decrease in e. Marginal cooperation costs are also the function of e, which decreases with the increase in e. In an anarchy governance structure, participants in different regions each holds a fairly large amount of residual control power and there are, therefore, very big opportunities exist for resorting to opportunism behavior and the cooperation costs are high. If the cooperation costs of collective actions are too high for the society to pay, the only way to avoid their conflicts is to increase the levels of the governance structure hierarchy so as to reduce the possibilities of opportunism and the cooperation costs.

Although cooperation cost is regarded as the function of governance structure, there are in fact many more kinds of costs that affect cooperation. The first is the characteristics of the participants. Generally speaking, the stronger the asymmetry among participants, the higher the cooperation costs and vice versa. In water governance, due to asymmetry of negotiating forces among different regions, their concerted action reached is naturally concomitant with very high cooperation costs. If participants can easily turn to other cooperation partners, opportunism would lead to less losses and lower cooperation costs. Otherwise, if cooperation partners are fixed, the opportunism of cooperation partners would lead to bigger losses and higher cooperation costs. The cooperation partners are usually fixed in water governance and that has boosted the cooperation costs.

Secondly, the asset specificity of the target of cooperation affects cooperation costs. The higher the asset specificity of the goods involved in cooperation, the higher the losses that would be caused by opportunism of cooperation partners as in the case of having only water source in a region on the lower reaches of a river. In order to prevent the occurrence of such opportunism, the region has to make extra investments, which refers to a higher cooperation costs caused by asset specificity. This is universal in water governance. Even under the condition in which all participants fully honor their contracts, due to the incompleteness of contacts, opportunism may occur under many uncertain situations. Thus, the stronger the uncertainty of the environment, the higher the possibility of the occurrence of opportunism and the higher the cooperation costs. In the circumstance there is a strong variation of water, it is very costly to seek collective action by horizontal cooperation. The changing circumstances are an important factor for increasing the levels of the governance structure hierarchy.

Many other variables can also affect cooperation costs. For instance, a social circumstance or cultural tradition full of cooperation spirit is conducive to lowering cooperation costs. Otherwise, it increases such costs. The greater the number of participants, the higher the cooperation costs and vice versa. The more complicated the water systems, the higher the cooperation costs, and vice versa. In the framework provided by the model, factors affecting the cooperation costs change the inclination of the marginal cooperation cost curve in Fig. 2.5.

2.4.2 Management Costs

The rise in the governance structure hierarchy will increase management costs. Management costs come from the principal and agency relations between the decision-making entities at superordinate and subordinate levels. The objective functions of decision-making entities at a lower level and the principal are different as they would not carry out superordinate orders unconditionally. The incentives for agents to carry out superordinate orders are often inadequate. They tend to distort information in their own interests so that the efficiency of the structure drops. In order to prevent opportunism of agents, the superordinate decision-making entities must commit resources to control, oversee and coordinate subordinate agents so that their behaviors would get closer to the objectives set by the principal.

Our model also regards management costs as the function of governance structure e, which increases with the increase in e. Marginal cooperation costs are also the function of e, which decreases with the decrease in the value of e. As the price of reduction in cooperation costs, the management costs will increase with the levels increased in the governance structure hierarchy. This is a force that restricts the infinite rise in the hierarchy. The main reason for the increase in management costs is the reduction in the residual control power in the hands of subordinate decision-making entities, and with the rise in the degree of hierarchy its own incentive drops accordingly, thus leading to more outstanding problems between agents and the principal. The lower the efficiency of the hierarchical governance structure, the higher the management costs rise and the bigger the inclination of the marginal management cost curve.

Like cooperation costs, there are many other factors influencing the management costs. Usually, the greater the number of agents, the longer the agency chain, and the greater the asymmetry of information between the agents and the principal and the higher the management costs, and vice versa. Information asymmetry is one of the root causes of agency costs. The greater the asymmetry of information, the higher the management costs and vice versa. The role of factors affecting management costs is to change the inclination of the marginal management costs in Fig. 2.5.

2.4.3   Equilibrium in the Governance Structure

The optimal governance structure is a structure that has the least transaction costs. When marginal cooperation costs are equal to the marginal management costs, the governance structure is the equilibrium structure with the least transaction costs. To put it more simply, the option for water governance structure is, in reality, the trade-off of advantages and disadvantages brought about by the rise in the hierarchy structure. The main advantages of the hierarchical structure are: (1) it can obtain greater security guarantee, which is exhibited in the mobilization and regulation of resources by the authoritative decision-making entities in the environment of great uncertainties; (2) it can obtain the greatest efficiency in the unified planning for water governance; (3) it can provide more coercive implementation mechanism in settling conflicts of interests. The main disadvantages are: (1) the concentration of residual control power and the arbitration of the superordinates in decision making would result in the increase in the possibilities of errors; (2) it would reduce the incentives to decision-making entities at lower levels and aggravate the distortion of information in the hierarchy; (3) the authority derived from hierarchy is easy to make the whole system rigid and slow in response. In a word, the advantages of the hierarchical organizations lie in a high level of security while the disadvantages lie in a low level of efficiency. The anarchical organization can fully mobilize the initiatives of all quarters, however, it is very difficult for it to mediate conflicts.

The cooperation cost curve and the management cost curve go together to determine the equilibrium governance structure. Statically, if the marginal curve of cooperation and management costs can largely be fixed, the point where the two curves cross is the equilibrium of the governance structure. Dynamically, with the changes in the influencing factors, the forms of the two curves may all change, thus causing the marginal curves and the equilibrium point to move. For instance, if the uncertainties and sudden changes increase in the hydrological system, they would raise the cooperation and management costs, with the cooperation costs rising even faster, thus the equilibrium point will move toward higher tiers of the hierarchy structure. If there is progress in hydrology and information technology, the cooperation and management costs would be lowered, with the management costs lowering faster, thus increasing the levels of the governance structure hierarchy.

2.5   Origin of China's Hierarchical Water Governance Structure

2.5.1   Views of the Water Governance School

The birth of China's hierarchical water governance structure is a response to the frequent droughts and floods and the consequences of the interaction between human and nature. The hierarchical structure took shape in the year 221 when the Qin Dynasty unified the country. The national unification means the establishment of the hierarchical water governance structure. In fact, hierarchy is not only a feature of water governance structure but also a feature of social governance structure. The origin of the water governance hierarchy and the origin of the unified political system have inherently close relations. The following is a review of the origin of this idea.

There are many scholars who have touched the subject over the past century, that is, the inherent relations between China's early unification and the unique natural geography and climatic conditions. The "water governance" school represented by Marx and Wittfogel is most influential. The school holds that large irrigation projects have their unique importance in the East and they are of great significance in the emergence of the oriental despotism or centralism (Curtis, 2009).

Marx was the earliest to see the importance of irrigation project in the Asian civilization. In the Orient where civilization was too low and the territorial extent too vast to call into life voluntary association, the interference of the centralizing power of Government. Hence an economical function devolved upon all Asiatic Governments, the function of providing public works. This artificial fertilization of the soil, dependent on a Central Government, and immediately decaying with the neglect of irrigation and drainage, explains the otherwise strange fact that we now find whole territories barren and desert that were once brilliantly cultivated, as Palmyra, Petra, the ruins in Yemen, and large provinces of Egypt, Persia, and Hindostan; it also explains how a single war of devastation has been able to depopulate a country for centuries, and to strip it of all its civilization (Marx & Engels, 1972, p. 64). Wittfogel goes a step further when he points out in his "*Oriental Despotism*" (Wittfogel, 1957) that irrigation was the primary cause for despotism in China. It is defined that civilizations whose agriculture is dependent upon large-scale waterworks for irrigation and flood control are 'hydraulic society', which featured despotism and centralized bureaucratic administrative system. Although Wittfogel's theories of are much disputed, they provide clues to the importance of irrigation in the oriental civilizations (Gregg, 2016).

Huang (2002) further developes the theories of the water governance school. He holds that the unification of China was a response to the force of nature and irrigation played a major role. Relative to the irrigation projects Marx and Wittfogel stressed, Huang puts more stress on the importance of fighting against floods. He gives special stress to the severity of the floods of the Yellow River. He argues that local control would come to nowhere. Only there is a centralized government that controls all the resources and gives equal treatment to all parties concerned, is it possible to ensure security and free the people from constant threat. It is underlined that water control alone is enough to demonstrate that the centralized system is inevitable in China. Moreover, regarding unification of China, Huang believes that the Yellow River flood is an important variable, but not the only one explanatory variable. It suggests that work relief and national defense are the two factors for promoting the unification of the country (Huang, 2002, pp. 6–10). All the three factors are determined by the natural geography of China and that is why Huang holds on to the view that natural force is the more decisive factor behind the unification of China. Toynbee advances the 'challenge and response' model for the rise of civilization. The ancient Chinese civilization was originated in response to the challenges by the difficult conditions of the Yellow River just as the ancient Egyptian civilization was originated in response to the challenges by shrubs and marshland in the Nile Delta and the Maya civilization was originated in response to the challenges by the tropical forests and Minoan Civilization was originated in response to the challenges by the sea (Toynbee, 1989; Lang, 2011). Needham (1981) sees the importance of water control in China, saying that the existence of the 'bureaucratic system' is the need of safeguarding the irrigation system.

2.5.2 Particularities of China's Option for Water Governance Structure

The preceding section advances the choice model for governance structure, which may provide scholars of the water governance school with an official theoretical connotation. This section explain why China has opted for the hierarchical water governance structure and, in the next section, it will show how the hierarchical water governance structure led to the hierarchical structure in state governance. Based on our model, the water governance structure of a country is determined by the comparison of cooperation and management costs. The particularities of China's option for water governance structure lies in the high cooperation costs, which is expressed in the difficulty for carrying out equal and cooperative collective action in the supply of large scaled public services such as flood prevention and work relief.

Now, first of all, let us examine the option for collective action in fighting against floods. China's unique monsoon climate and geographical conditions determine the big threat of floods, especially from the Yellow River, because it carries the largest amount of mud and sands downstream (Xia & Pahl-Wostl, 2012). The river course on the lower reaches is very easy to be choked. Although large scaled collective action took place during the Qin Dynasty when the social productivity was at a very low level, the people were gravely threatened by floods. Without a highly authoritative organization to coordinate the actions of various regions, the local conflicts were constant, as cooperative action was quite difficult. This is evidenced by historical literature. In the Western Zhou period, dykes already appeared on the lower reaches of the Yellow River, indicating the need for collective action among various regions. During the Spring and Autumn period, various kingdoms on the lower reaches began to build dykes, which threatened each other's security. Some kingdoms viciously imposed disasters on neighbors. In order to prevent the act of "making neighbors as water outlet" and "using water as a weapon", there was the need to make contracts among the minor kingdoms. In 651 B.C., Qi Huan Gong called all the dukes to meet in Kuiqiu and took oath not to build dykes secretly and ban secret dykes. This shows that the kingdoms of dukes ensured each other's security against floods by signing agreements. Up to the Warring States period, the dykes on the lower reaches of the Yellow River began to assume a considerable scale, which is evidenced by the description by Jia Rang of the Western Han Dynasty, which states that the building of river dykes started in the Warring States period to protect all the rivers and all benefited from them. The State of Qi shared a river as borders with the State of Zhao and the State of Wei, which set against mountains. The State of Qi built a dyke along the river, extending 12.5 km. The river flowed eastward to the dykes of the State of Qi. Threatened by floods in the west, the State of Zhao and the State of Wei also built dykes, extending 12.5 km (Waley, 2012). The State of Qi was the first to build a dyke due to its low-lying terrain to ward off floods. Seeing that the floods invaded Zhao and Wei, the two states also built dykes. In 332 B.C., the State of Zhao was engaged in a war with the State of Qi and the State of Wei and breached the Yellow River dyke to flood them. This shows that the states of dukes could not reach unanimity in collective action by taking oath and it is impossible to ensure security against floods by signing agreements. After the Qin unified the country, water governance, which was cut up by various states during the Warring States period, was also unified, with all the military installations that obstructed water flow and all the passes that obstructed communications removed, making it possible to link up all the states of dykes. This is what was done by Qin Shi Huang, the first emperor of the Qin Dynasty (Sima, 2007).

The difficulty in cooperation on an equal footing is seen not only in the flood control of the Yellow River but also in disaster relief, including both droughts and floods. As natural disasters visited all different states of dukes every year, there was a need of "mutual assurance" and sharing of the obligations of disaster relief among the states. When there was a famine and neighboring states fail to provide relief, it would probably lead to war. At the Kuiqiu meeting, the participating states also promised not to wage war in famine. However, such promise was in fact very feeble. War took place very often. According to the records of "Zuo Zhuan", famine struck in the State of Jin in 647 B.C.. The State of Qin came to its rescue. The following year, famine visited the State of Qin. Yet, the State of Jin, instead of providing relief in return, waged a war against it. There were many wars like this during the Spring and Autumn period. Huang Renyu points out that the above background could only increase the possibilities of conflicts. Vexation by war for hundreds of years gave rise to the desire for unification. The benefit of unification is that a big state controls more resources and can more efficiently provide relief to disaster victims (Huang, 2002).

The particularities of China's choice for water governance structure lies in the changing natural and geographical environments. In the early period of civilization, there was the need for large scaled collective action in controlling floods. But under the then productivity level, the costs for collective action based on equality among states were extremely high. In our model, it means the marginal cooperation cost curve MC_n is very steep, resulting in a very high degree of the equilibrium water governance structure hierarchy e^* (see Fig. 2.6). Our model shows that the

Fig. 2.6 Economic explanation of China's choice for water governance structure

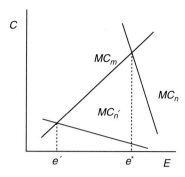

hierarchy structure is optimal for China in water governance. In the early period of the growth of the Western civilization, for instance, in the ancient European continent, there were neither so capricious weather conditions as in China nor such disastrous floods like those brought by the Yellow River. There was, therefore, not the need for large scaled collective action in water control. That means that under the natural geographical conditions in Europe, the cooperation costs needed in collective action in the public service like water governance among countries was quite small, which is shown as very flat in the marginal cooperation costs curve MC'_n, hence the very low level of the water governance structure hierarchy e'. (see Fig. 2.6). The comparison is conducive to our understanding why China needed a highly centralized and unified system in water control—the cooperation costs in joint water control were too high, that is, the cost of political negotiation mechanism among states was too high. The highly centralized system can economize the cost of political negotiation to the maximum. It is, therefore, an equilibrium governance structure.

2.5.3   The Formation of Hierarchical State Governance Structure in China

According to the theories of the water governance school, the requirements of water governance for centralized system are the inherent driving force behind the unification of China. In the dissertation language, it means that the demand of water governance for hierarchy structure determines the demand for the hierarchy structure in the governance of the state, ultimately leading to the birth of an empire. Starting from Western Zhou period, there were more than 100 states ruled by dukes. The water governance structure was also dispersed. That is a kind of instable governance structure. The society thus developed an internal driving force for contraction toward equilibrium structure, thus leading to, step by step, great unity from the segmentation by dukes. This is a process of long-time games, a process of transition from instable governance structure to stable structure. Before Qin unified the country, China had been in a long period of mutual slaughter. The original more than 100 states were reduced to dozens and ultimate to 13. In the last 200 years of the period there were only seven bigger states left. Qin annexed the other six to complete the unification. There has no such a scaled centralized movement in the world history. It is no doubt that natural forces are the deciding factors behind the unification of the country (Sima, 2007).

In the framework provided by our model, China, which used to be cut up by more than 100 states, became one country after more than 500 years of merger and what is behind is the natural force and the too high cooperation costs in the supply of public affairs that led to prolonged wars and conflicts. The essence of the matter is the extreme instability of the flat governance structure. Such a society has the powerful internal driving force to lower the governance costs by raising the flat governance structure to a high degree of hierarchical structure and at the end it is evolved into a stable hierarchical structure. This vibrating process that extended for as long as 500 years as shown in $e^0 \rightarrow e^*$ in Fig. 2.7, is a process of movement from a point far away from the equilibrium point to the most optimal equilibrium point, thus giving rise to a hierarchical structure in the governance of state, that is the empire system.[5] This shows that the formation of the hierarchical governance structure is, in fact, for the purpose of economizing cooperation costs, that is, the cost of political negotiation in the inter-regional public affairs. However, while economizing cooperation costs, the hierarchical structure has to pay a fairly high price for management in order to maintain the stability of the governance structure, hence there is its internal driving force for lowering management costs. Once the hierarchical structure is formed, it would develop its culture compatible with the system, lower management costs, thus enabling the marginal management cost curve to move inwardly, which is shown in $e^* \rightarrow e^{**}$ in Fig. 2.7, a movement toward the higher and more optimal equilibrium point in the hierarchy, the higher the degree of hierarchy, the more centralized power in the empire. The above

Fig. 2.7 Early origin of China's hierarchical water governance structure

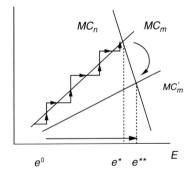

analysis reveals two features of the hierarchical structure: first, the formation of the hierarchical structure is the result of economizing cooperation costs; second, the maintenance of the hierarchical structure necessitates the lowering of management costs, hence determining the "self-reinforcement mechanism" of the hierarchical structure. In other words, once the centralized structure is formed, the degree of centralization would tend to be self-maintaining and reinforcing. The logic is applicable to the hierarchical water governance structure and also to the unified empire system.

The above is a re-explanation of the theories of the "water governance" school by employing the model constructed in this book. It is not a simple repetition of the existing thought but new propositions deduced from the existing theories by employing a modern economics model, which have enriched and developed the existing theories. The following are the three major propositions that have brought the theories of the "water governance" school into depth.

First, China's unified system is the product of transaction costs (cooperation costs) economizing. The book analyzes the logic that the building of an empire is for the purpose of economizing the high cooperation costs by replacing vertical administrative control with horizontal political deals. This almost runs in the same groove as Coase Theory of the Firm. This discovery is the deepening of the theories of the "water governance" school, which holds that the unified system is the demand for centralization of power in water control. This book has further revealed that the exact meaning of such demand is to lower cooperation costs in the supply of inter-regional public affairs. The unified system is, in essence, a model of collective action by exercising the centralized power and vertical orders that replaces the model of collective action featuring separation of power and horizontal association. This is the result of the efforts by the early civilization to cope with the challenges by nature and seek transaction cost economizing.

Second, a unified empire has the powerful motivation to lower transaction costs (management costs). In order to maintain its stability, an empire has to lower effectively management costs and make institutional arrangements, cultural orientation and ideology compatible with the centralized system. This can explain the series of important measures adopted by all kingdoms in a bid to strengthen the centralized system. For instance, after unifying China, the Qin Dynasty replaced the feudalism system with the "*Jun (province)-Xian (county) system*", unified measurement system and the language, burnt books and buries alive Confucian scholars and banned freedom of thought. All these institutional arrangements were aimed to lower management costs. Later dynasties drew on the experience of their predecessors to introduce new institutional arrangements to further lower management costs. The Western Han Dynasty introduced the policy of "*banning all schools of thought but Confucianism*" to lay the cultural foundation compatible with the unity of the country. Sui and Tang dynasties introduced the imperial examination system to select officials as spokesmen for the ruler. These major institutional arrangements constantly reinforced the empire system.

Third, the disintegration of an empire is the result of inability to effectively control transaction costs (management and cooperation costs). Effective lowering of management cost is the prerequisite for maintaining an empire. Management costs are mainly the costs of entrusting agents, that is, the costs for the operation of the entire bureaucratic system. At the beginning of a kingdom, as immediate interest groups were smashed or weakened, it was easy for the ruler to lower management costs and the operation of the bureaucratic system was fairly efficient. But, with the passing of the time, new interest groups began to form and the operation of the bureaucratic system became less efficient and even rigidified, with corruption and bad administration running wild, thus resulting in a rapid rise in management costs. When the management costs rose too high, it became difficult for the centralized governance structure to go on, thus causing the social cooperation costs to rise rapidly. All these added to the total transaction costs until the society was unable to bear and the empire was going to collapse. But as the motivation for national unity was still there, and there was the inertia of civilization, that is, the features of 'path dependence' (David, 2007) for the changes of system, there was still the tendency of re-embarking on unity after the separation. This is, in reality, an explanation of the changes of ancient kingdoms in China.

2.6 Water Governance Structure and State Governance Structure

There are inherent relations between the water governance structure and the state governance structure, because water is a public resource involving the interests of all people and the population covered by water governance and state governance is the same. For most countries, water governance structure is selected after the state governance structure takes shape. State governance structure is exogenous of water governance structure and has a major impact on the option for water governance structure. In the early period of the Chinese civilization, due to the particularities of water problem, such as frequent droughts and floods and the severity of the Yellow River floods, the option for water governance structure had a decisive impact on the option for the state governance structure.

Why, then, water governance has such an explanatory power of the Chinese civilization? The following is a further illustration provided on the basis of Fig. 2.3. In this model, we have added another new conceptual curve, known as the "minimum governance ability demand curve", which refers to the minimum governance ability required for a certain degree of water security in an area. It is expressed in a dotted production possibility curve in Fig. 2.8. We assume that it is relative stable. The particularity of China is that in ancient China when productivity was at a very low level, the maximum possible governance ability curve(C_0) was still lower than the minimum governance ability demand curve (D). That means that even in the early period of the Chinese civilization, the country was gravely challenged by water. Under the low production and technology conditions, even if the country acquired the maximum possible governance ability, it was not enough to ensure

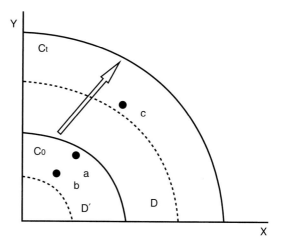

Fig. 2.8 Minimum governance ability demand curve and water governance structure options

water security. In face of such tremendous pressure brought about by the threat of floods, the society had the powerful intrinsic driving force to opt for a governance structure that could economize transaction costs to the maximum, thus making the actual governance ability to get closer to the maximum to the then maximum governance possibility curve (point a). Such governance structure is a highly centralized hierarchical structure, which is the most optimal to cope with the challenges by nature under the then low level of productivity.

The civilizations in Europe and the Americas were quite different from the Chinese civilization. In the early period of the European and American civilizations, challenges by floods were not so grave to the society. The minimum governance ability demand curve is not only at a very low level (D′) but also far lower than the then maximum governance possibility curve (C_0). It was not necessary for such society to tap fully the maximum governance ability. A quite low actual governance ability (point b) is enough to satisfy the need for ensuring water security. The early civilization was in such an environment that water governance did not have much explanatory power about the growth of civilization. But just the contrary is true with China. In such a civilization, state governance structure determines water governance structure and the formation of state governance structure requires explanation by other factors. The comparison of the Chinese and Western civilizations shows that a hydraulic society is indeed an important clue for knowing the unique features of the Chinese civilization.

In the ancient agricultural society after the Qin Dynasty unified China, water governance still had a strong explanatory power about the Chinese empire that continued for more than 2000 years. It is discovered that in the time-honored ancient Chinese history, even in the peaceful period, the country did not have enough strength to resist the Yellow River floods. During the 23 years of the Reign of Zhen Guan of the Tang Dynasty, the Yellow River was flooded in eight; in the 134 years of the reign of Emperors Kangxi and Qian Long of the Qing Dynasty, the Yellow River was flooded in 47 (Zhu, 2004). This shows that under the production and technology conditions of the ancient agricultural society, the economic surplus and technical ability accumulated by the society were hardly enough to ensure water security in the Yellow River basin. It can also be deduced that the maximum governance possibility curve (C0) was always below the minimum governance ability demand curve (D). Till the modern society, due to the development of industrial civilization, the economic surplus has increased in large amounts and technology has progressed rapidly, making the maximum governance possibility curve move far away in the outward direction (C_t), higher than the minimum governance ability demand curve (D). That is the basic pre-condition for ensuring the security of the Yellow River and realizing peace every year. Although the hierarchical governance structure is also the institutional guarantee against Yellow River floods, because the maximum governance possibility curve is already higher than the minimum governance ability demand curve, it is not necessary to make the actual governance ability to get as close as possible to the maximum governance possibility curve. What is required is to make it slightly higher than the minimum governance ability demand curve (point c). We can also say that the demand of the modern society for the hierarchical water governance structure has dropped to a certain extent and so has the explanatory power of water governance structure on the state governance.

The unified system of China is super stable, just as the "Qin system" that lasted for thousands of years. Although ancient China saw the successive rules by dozens of dynasties, the centralized system as the basic form of state governance has not changed much.[6] Such super stable structure is closely associated with the hierarchical governance structure, which is the result of the response to the challenges by nature. Once it was formed, the centralized empire would by all indigenous means to maintain the central power, thus giving rise to a political structure and economic system compatible with the centralized system and also giving shape to the unified culture and mentality. These and the natural geographical conditions have made the centralized system acquire its self-maintenance inertia and shaped up a form of civilization that can constantly strengthen itself. Such civilization has a tremendous inertia. It is impossible to reshape fundamentally the civilization without revolutionary external forces. China is, therefore, a 'hydraulic society' for thousands of years. Even today, it is left with the lasting imprint of a hydraulic society.

---

[6]In the more than 2000 years since the Qin Dynasty, China was in unity for two-thirds of the time and in disunity in one-third of the time. Even during the period of disunity, the popular feelings also pointed to unity. So, unity is the constant state of China. Such super stable centralized system that has continued for more than 2000 years is unique in the world history.

Chapter 3
Hierarchical Structure of Water Rights

This chapter constructs a model for studying water rights on the basis of an indepth discussion about the term "water rights". The structure of property rights in water resources is very complicated, far different from ordinary economic assets. This chapter reviews the development of theories on property right structure of natural resources in the West and it discovers a "conceptual model of institutional hierarchy", which is applicable in analyzing the property right structure of flowing natural resources, being the foundation of developing a "conceptual model of water rights hierarchy" for understanding the water rights structure in China. The chapter will give particular stress to the differences between China and the Western societies in terms of water governance structure and the water rights structure resulting from the differences which cannot be reflected in descriptive models. China's water rights structure is a special hierarchy, which has absolute precedence of public rights over private rights, with the latter greatly attenuated. There are also the features of a series of other rights compatible with the hierarchical structure.

3.1 The Meaning of Water Rights

This book touches on the basic concept of "water rights" in its introduction and defines it as "water rights in the broad sense" and "water rights in the narrow sense". The water rights examined in this book are water rights in the narrow sense, that is, property rights in water resource or property rights in the use of a given amount of water under condition of scarcity.

The object of water rights is water resource. According to property rights economics, property, as the object of property rights, has to satisfy three conditions: First, it must have utility value, that is, it must be useful. Useless things are not property; second, it must be possessed, that is, it must be controlled and used; and third, it must be scarce. By these standards, water resource must have three requirements to become property: utility, controllability and scarcity. The precondition of

© Springer Nature Singapore Pte Ltd. 2018
Y. Wang, *Assessing Water Rights in China*, Water Resources Development and Management, DOI 10.1007/978-981-10-5083-1_3

utility is that water may be used when needed. For instance, the farmland at a given period of time in a given area is dry and it is beneficial to irrigate the land during this period. But after a rain, irrigation water might become superfluous. Another example is: the water flow in the upper reaches of a river is dammed for storing water, causing water shortages in the lower reaches. In normal conditions, the water dammed up is useful in both upper and lower reaches. But flood water in either upper reaches or lower reaches is useless. So, the utility of water resource is built on the basis of controllability. Water resource in its natural state is hard for mankind to utilize on a large scale. Only by engineering measures such as building flood storage projects, weirs and water diversion projects, is it possible to satisfy water demand in both time and space, ensuring that when water in need is available for use. With the social and economic development and technical progress, more and more water resources have been brought under control and utilization. Such non-conventional water sources, such as seawater, slightly salty water and sewage water were not used in ancient times. But now, engineering measures can put them to use. Even flood water can be used as a resource when the ability of flood regulation and storage rises in modern society. If water resources are abundantly available, people can obtain it according to needs and there is not the problem of property rights. As water resource scarcity is the precondition for talking about water rights, it is necessary to distinguish the two kinds of reasons for causing the scarcity: scarcity due to lack of engineering projects and scarcity due to water availability. The former refers to inadequate input into opening up, economizing or protecting water resources, when the potential for water supply can be tapped by increasing input. Water shortage resulting from inadequate input into engineering projects or from pollution is regarded as water scarcity due to inadequate engineering projects. When input into engineering projects increases and the potential is exhausted, water scarcity is mainly caused by limited water availability (Edelenbos, Meerkerk, & Van Leeuwen, 2015). Theoretically, if input into engineering projects increases infinitely, resource scarcity can always be eased by cross-basin water regulation, progress of technology for water economizing and non-conventional water sources development. But as capital input is limited, it is necessary to study the feasibilities for developing and utilizing water resources. So water scarcity due to inadequate engineering projects and water scarcity due to inadequate water availability are relative.

Generally speaking, the object of water rights is the water resource that can be and is likely to be used under the existing technical and economic conditions. It must be scarce in terms of quantity, whether due to inadequate engineering projects or due to limited availability. What needs pointing out is that the object of water rights in the broad sense is much broader than that of water rights in the narrow sense. As long as there is water control, such as engineering projects, anti-flood projects and waterway shipping, there is the problem of water rights in the broad sense. Whereas water rights in the narrow sense is meaningful only in its beneficial use and allocation when water resources are scarce. In reality, "water rights", a new term in China, has been raised against the background of the daily aggravating water shortages.

In the following, we shall examine the contents of water rights. Property rights, in essence, are certain rights to resources but not the resources per se. For simple resources, the rights and resources per se are integrated. But for complicated resources, the rights and resources per se are separated. Water resources belong to the latter. Water resources are flowing and can be recycled for use. The object of water rights is perceptually a certain amount of water and is, in essence, a bundle of rights on the beneficial use of a certain volume of water, which may be termed as a bundle of water rights. A bundle of water rights can be divided into many types. This book divides the water rights bundle into the right of allocation, the right to withdraw and the right to use. As all the rights in the bundle can be separated, the term water rights can refer to a single right in the bundle or used in plural, as the whole of the right bundle.

What needs special explanation is the allocation right in the water rights bundle. The right to allocation of water resources is political right and also a right to property, with the administrative management right and property right overlapped. Theoretically, all resources within the boundary of a country are controlled by the sovereign state. Even those countries with a private property right system, the state still has the right to plan, oversee and manage resources in the public interests and levy taxes on resources development. In a hydraulic society, the government exercises the public management function and the function of owning the rights to resources property, thus subordinating property rights to administrative rights. On the other hand, the overlapping of administrative and property rights is, to a large extent, associated with the collective action in the allocation of water resources, which is realized through political power exercised directly by the government in a hydraulic society. So, the overlapping of property right and administrative right is easy to understand.

In a word, water rights are rights to water that can be put to beneficial use when the water resources are scarce. They may be regarded as an exclusive decision making right by a given decision making entity on the given amount of water it holds. It includes two aspects: first, the decision-making right concerns the right of allocation, the right to withdraw and the right to use, which are exclusive. Second, the amount of water resources allocable under the decision making right, that is, the volume of water to be allocated, withdrawn or used. Due to uncertainties in the hydrological features, the quantity of water corresponding to the water rights is fluctuating in terms of time and so will be the quantity of water allocated for beneficial use. In practice, it needs dynamic regulation. In this regard, I recommend that 'water rights quotas' or 'water rights limit' represents the quantity of water corresponding to the water rights.

The right to allocation, withdrawal and use of water can also be subdivided into two parts: public right, referred to as right to allocation, and private right, referred to as right to withdraw and use. The former is a political right and the latter, an economic right. The political right exercised by the state is, in fact, part of the property right, an overlap of administrative and property rights. This is determined by the public nature of water resources. From ancient times to the present, China is a society with a public ownership of water rights. The political right to water exercised by the state in the name of the public has been stressed again and again and the private benefit from water is put within the public interests and the private rights to beneficial use has long been attenuated. Water rights in the sense of property have long been ignored. It is even more so during the period under the planned economy. Although water rights, as objective rights, have been always there, China has long avoided the use of the term 'water rights'.

The changes in the social background since the country introduced the Reform and Opening-Up Policy have stimulated greater demand for the property rights to water resources: on the one hand, economic development has rapidly expanded the scope of water scarcity, resulting in a strong competition among different regions, departments and groups in the use of water and augmenting the attributes of property in water resources; on the other hand, the market economy has given rise to a large number of independent interest groups, property rights owners and investors, who have raised their demand for property rights to water resources. In this context, it is not enough to stress the political rights of the state to the use of water in settling water conflicts, raising the efficiency of water use, optimizing the allocation of water resources and absorbing investment in water resource development. It inevitably requires greater respect for property rights to water resources. The concern about 'water rights' is therefore inevitable.

3.2   Review of the Natural Resources Property Right Theories

We have clarified the concept of water rights and now let us examine the structure of the property rights of water resources, which is the sum total of rights and obligations of all decision-making entities concerning the use of water or which may be construed as the form for realizing the property rights to water resources or the state of distribution of such resources in the real world. The structure of water resources property right is complicated, which is diversely different from ordinary economic assets.

3.2.1   Dichotomous Classification of Property Rights

Since Adam Smith, the mainstream economics has advocated for the property right structure that is the most favorable for the growth of national wealth, that is, private property corresponding to the market mechanism. Private property rights refer to the bundle of rights owned exclusively by a given entity, which include the right to use, the right to transfer and the right to capture income or benefit. The complete private property rights refer to the right to use, the right to income and the right to free transfer that are exclusively possessed by an entity. But the actual property rights are often constrained, thus making private property rights attenuated.

The property owned by multiple parties is called common property. In the early economics literature, all non-private property is defined as common property, which is often used to refer to the characteristics of the property rights of natural resources. The most usual classification of property rights is a distinction between private property and common property (Rymes & Gordon, 1991). It is discovered that fishing resources are often destroyed and overused due to the characteristics of common property, a phenomenon called the 'tragedy of the commons' (Hardin, 1968). This vivid analogy reveals that when a scarce resource is owned commonly by many individuals at any time, it would be destroyed and the environment would deteriorate.

The tragedy of commons does not only take place in marine fishing grounds and collective pastures but also around us, such as collective forest farms and biological resources, acid rains, river pollution and flow cut-offs in sections of river courses. Modern natural resources economics has arrived at the conclusion after comparative analysis that so long as public resources are open to a number of people, the total extraction of the resources would be far more than the optimal extraction level (Ostrom, 1990). Economics also associates the 'tragedy of commons' with market failure, because it is difficult to establish a perfect exclusive private property rights to natural resources and therefore it is impossible to allocate the resources by making full use of the market mechanism (Francisco & Jorge, 2011).

3.2.2   More Detailed Classification of Property Rights

From the 1970s to 1980s, the dichotomous classification was criticized by many scholars due to the failure to recognize property rights exercised by governments and collectively by finite groups of people (Bromley & Pearce, 1992; McDaniel, 2001; White & Costello, 2011). This resulted in a widely accepted description of four classes of property right regimes as described by Bromley (1989), that is, state property, private property, common property and non-property (open access) in terms of the holders of rights and the corresponding duties of the rights holders and other parties. Such classification by Bromley is shown in Table 3.1:

Even this classification is a broad definition of property rights. More scholars have become aware of the fact that the property rights in realities might be continuous instead of such dispersion as dichotomous or four class regime. Private property and common property are the two ends of property right arrangements, with most property rights in between (Cheung, 2000). It is argued that property right structures of natural resources as a continuum, emphasizing that natural resources

Table 3.1 Four classes of property right regimes

	State property	Common property	Private property	Open access
User limitation	As determined by state agency	Finite and exclusive group	One person	Open to anyone
Use limitation	Rules determined by a state agency	Rules determined by mutual agreement	Individual decision	Unlimited

can by and large be classified as exclusive assets and common assets (open access resources assets) according to the exclusive use. In fact, a continuum is formed from exclusive assets and public assets, with the former including land, solid minerals and forests and the latter including marine resources, flowing water bodies and environmental bearing capacity (Aubin & Varone, 2013; Schmidt & Mitchell, 2014). Even so, the four classes of property right regime have been widely applied for its reasonable abstraction of the practical arrangements.

3.2.3   Common Property Rights Structure Closer to the Real World

Since the 1990s, scholars have further recognized the complexity of property right structure. For instance, the offshore fishing resources are generally common resources of local fishermen, but all countries of the world and fishermen of different areas are widely divided in the characteristics of rights. Edella Schlager and Elinor Ostrom have studied the disparities and offered their own version for the classification of common property. Such classification shown in Table 3.2 has been held for a time as the prevailing classification of common property rights (Schlager & Ostrom, 1993).

Edella Schlager and Elinor Ostrom (1993) distill the bundle of property rights into five categories: access, withdrawal, management, exclusion and alienation, saying that actors in most cases only possess part of the property rights. According to the rights possessed, they classify actors into four categories: owners who possess all the rights, proprietors who do not possess the alienation right, claimants who only have management, access and withdrawal rights, and authorized users who only have access and withdrawal rights. In their studies of coastal fisheries, they provide many actual cases to support the five categories of property rights based on the above framework. Their studies have further revealed the diversity and complexity of property rights.

Ostrom and her colleagues then go on to advance a multiple-tiered analysis method. In her renowned work *"Governing the Commons"* (Ostrom, 1990), the multi-tiered analysis is introduced into the rules of use of common pool resources. She advances the rules for determining the property rights at the three levels: constitutional rules, collective rules and operational rules. All acts happen at the

Table 3.2 Common property right classification by Edella Schlager and Elinor Ostrom

Rights bundle	Property rights holders			
	Owners	Proprietors	Claimants	Authorized users
Access and Withdrawal	√	√	√	√
Management	√	√	√	
Exclusion	√	√		
Alienation	√			

nested levels in the use of common pool resources and the changes in the rules at one level is restricted by the rules at the next higher level, which go to form a system of nested institutions.

3.2.4   *Institutional Hierarchy Model*

The multi-tiered property rights structure and the system of nested institutions revealed by Ostrom have had a big impact on the later researchers. On the basis of the works of Ostrom, a conceptual model of institutional hierarchy is advanced by Challen (2000) from the following four perspectives.

First, the nature of the decision-making entity holding the rights pertaining to use of a resource is made as the base for classifying the types of property rights rather than the nature of the entities holding the rights. Thus private property corresponds to a single decision-making entity, such as an individual person or firm; common property to a finite collective entity such as a cooperative group; state property to a government entity; and open access to the absence of any entity with decision-making power over a resource.

Second, for most resources there are multiple levels of property rights, starting with broad powers of state or national governments to control use of resources, and ending with powers of individual resource users to make investment and production decisions for resource harvesting and exploitation. In between these extremes may be more decision-making levels, all relating to some individual or collective entity with property rights over the resource. Parties at each level within a hierarchy have their own peculiar objectives in resource management and may make fundamentally different types of decisions, all of which ultimately produce a pattern of resource use.

Third, at each level of property rights, there are other types of institutions to manage natural resources. These were (1) entitlement systems that define the physical basis for dividing the resource amongst users; (2) mechanisms for making an initial allocation of entitlements amongst competing resource users; and (3) mechanism for making changes in the allocation of entitlements. Entitlements are re-allocated amongst holders of property rights according to changes in the socioeconomic or biophysical parameters of resource system. These three systems,

Table 3.3 Conceptual example of a property-right hierarchy in an international fishery

Scope of allocation problem	Parties to decision making	Conceptual property-right regime	Allocation decisions
Allocation of fish stocks amongst nations	Multiple national governments	Common property	Definition of territorial waters
Allocation of fish stocks amongst regional communities	National government	State property	Exclusive community rights to fishing areas
Allocation of fish stocks amongst individual fishermen	Community members or representatives	Common property	Individual transferable quota issued to fishermen
Allocation of quotas to fishing effort or sale to other fishermen	Individual fishermen	Private property	Private production and investment decisions

plus the decision-making entities, have formed the contents at the property right level.

Fourth, all the levels of property rights add up to form a hierarchy, which comprises a system of nested rules where each successive level is legally supported and maintained by the superordinate level.

Table 3.3 shows the application of property hierarchy in international fishery. The use and allocation of fishery resource have the participation of many groups. The property-right regime at any level in the hierarchy relates to the nature of the entity making the allocative decisions and is distinct from characteristics of the allocation decisions (Challen, 2000). Theoretically, this way of describing the hierarchy of natural resources is a big progress. Previous classification of property rights is a bundle of rights segmented at the same level by different entities. The hierarchy method has recognized that the bundle of rights is divided up vertically by entities of different levels that is especially useful in describing the characteristics of property rights to flowing natural resources.

In general, the classification of property rights in natural resources in the West experienced three development stages: the first stage is before the 1970s, which is marked by dichotomous classification of private and common property; the second stage spanning the 1970s to the 1980s, is marked by the four classes of property right regime, that is, state property, private property, common property and non-property (open access); the third stage is from the 1990s to the present, when the property right classification is more accurate and closer to realities, represented by the more realistic and more descriptive property right hierarchy.

3.3   Comments on the Theory of Property Right Hierarchy in Natural Resources

3.3.1   Academic Contributions of Property Right Hierarchy

Challen has received very high evaluation in the international academic circles for his study of natural resources property rights by employing the new institutional economics theories, as one of the representatives comparable with the classics of Ostrom. Property rights are of particular importance in the study of natural resources governance (Agrawal, 2007; Schroeder & Castillo, 2013), encompassing a diverse set of tenure rules, aspects of access and use of resources, and relationships between people (Meinzen-Dick, Brown, Feldstein, & Quisumbing, 1997). Challen has made major breakthroughs in the theories of natural resource property rights.

First, the nature of decision-making entities is defined as the basis for classifying property rights. The term decision-making entity is perhaps superfluous as far as simple property right structure is concerned, such as private property right, however, it is very useful to identify the nature of property rights in the complex property rights in realities, for instance natural resources. Challen identifies a property right holder by studying whether an entity has the decision-making power over the use of resources. This is clear and perceptual as compared with the traditional way of seeing the nature of the property right holders, which is viewed as a progress. The term 'decision-making power' means the right of choosing the way of how the resources are used, including multiple uses, the way the resources are used and the amount of resources is used. So long as there is the de facto right of choosing how to use resources, there is the de facto holding of the property rights. For instance, fishery resources are legally belonging to the state. But local authorities have the right to decide on how much is to develop, and how to allocate the resources and to whom they are allocated. Local fishermen have the right to decide on the way and quantity to catch. So local authorities and fishermen can both be regarded as de facto property right holders.

Second, it has recognized the complexity of natural resources, especially flowing natural resources (such as water), which is a hierarchy, an apt description of flowing natural resources that is closer to reality. Challen discovers that decision-making entities are distributed at all levels of the hierarchy, which have different objectives of resource use and make different decisions. For water resources, the entities at multiple levels from the state right down to individuals all have the power of controlling the use of water resources. That means that the rights of the bundle of property rights to water resources are held by multiple entities at multiple levels in the hierarchical structure. Before Challen, we could only say vaguely that the property rights to water resources are common property or a mixture of state property and private property. The hierarchical structure has opened the black box of the non-private property, making it possible to describe the common property more accurately.

Third, Challen has appropriately distinguished between property rights and institutions. He is against the proposition of equating property rights to institutions in the study of property rights to natural resources. He describes property rights as a subset of institutions and the nature of property rights is determined by the nature of decision-making entities in the use of resources. Moreover, he points out that at each level of property right hierarchy, apart from decision-making entities, there are other institutions using the resources, which include the entitlement systems, the mechanisms for making an initial allocation of entitlements amongst competing resources users and the mechanisms for making changes to the allocation of entitlements. These, together with decision-making, form a set of nested hierarchy. Although the institutions associated with resource use at each level are not limited to the above three, it is no doubt that he has listed the most important ones.

Fourth, Challen has made excellent economic explanation of the institutional hierarchy he advanced. He remarks that there are net benefits to be gained by retaining some allocative decisions at particular levels in a hierarchy and by delegating power to make other allocative decisions to subordinate and decentralized decision-making entities. He further comments that the logic for the choice of institutions at different levels is the same, that is, minimizing transaction costs, which may be used to explain the diversified institutional choice in reality. Apart from the explanation of static transaction costs, Challen has provided an explanation of dynamic transaction costs in the institutional hierarchy, thus unveiling the logic why the hierarchical structure changes according to the minimization of dynamic transaction costs. Obviously, the study by Challen (2000) is a successful paradigm in applying transaction costs in the natural resources and environment management, which may be regarded as one of the founding works of natural resources and environmental institutional economics.

3.3.2   Background and Limitations of the Property Right Hierarchy Theory

Challen's (2000) theoretical study is set against the background and empirical study of the Murray-Darling Basin, which is quite different from China in at least two aspects. One is that Australia is a federal state while China is a centralized state. The other is that the Murray-Darling Basin is in the southeastern part of Australia where there is vast land sparsely populated, with only three million people in the one million square meters of area. The flood problem is relatively less serious. While in China, the major river basins are densely populated and flood problem is very serious. The Yellow River Basin, for instance, covers less than 800,000 km^2 in area but it supplies water to more than 100 million people and the flood problem is extremely grave.

Water rights transfer in the Murray-Darling Basin started in 1984, within a small basin area at the beginning. It was spontaneous action of water users at the very beginning and the water market began to expand later on, when water trading spread to different small basin areas and different states. The government played a supporting and guiding role in the process. There have been a series of problems in water trading over the past 20 years in the Murray-Darling basin, which has given rise to the burning demand for theories about water rights and water market. It is in this context that Challen started the research project. He has given the following background that leads to his book.

In the mid-1990s, resource economists and policy makers showed great concern for private water rights and water rights trading. The Australian government began to define water rights and built a water market for re-allocation of water rights. But the institutional reform met with great setback in institutional designing and implementation. This shows that the water use problem could not be settled through private water rights. Water resources in a basin were regarded as big and complicated common pool resources. Reform involved a lot of collective action and externalities, thus restricting the effectiveness of private property rights and other existing property rights. Based on these problems, the study examined the introduction of private property rights to water resources and compares the economic advantages of different property rights, especially between common property and private property. Fairly soon, it was discovered that the choice resource allocative institutional structures are far more complicated than the simple property rights. The institutions for managing and using natural resources are very complicated, with the decision-making powers existing in many types and in large numbers and so are the decision-making entities. The optional institutions change among different right holders but not different in dispersion of property rights. In view of the complexity of resource reallocation, it is necessary to develop a conceptual model to explain and describe institutions.

'Hierarchy' in Challen's (2000) 'institutional hierarchy model' refers the multiple levels of property rights from the state down to individuals over flowing natural resources. This is might influenced by the new institutional economics (including the Theory of the Firm). However, Challen's institutional hierarchy does not have the basic characteristics of the hierarchy of the firm, such as strata, administrative control and coercive coordination. His original intention is not the hierarchy used in this book. Accurately speaking, what Challen refers to is a multiple level, as vertically overlapped structure. In the sense of new institutional economics and in its strict use of the word, Challen's institutional model does not conform to its name. It should be termed as an institutional bureaucracy model. It is the generic description of the bureaucratic system for managing natural resources (including property rights). Hierarchy is only one of the generic bureaucratic structures and very special one at that.

As Challen bases his study on the river basins of Australia, he could hardly have an insight into the real hierarchy that exists only in water governance in eastern countries. In the water governance continuum advanced in Chap. 2, the degree of governance structure hierarchy in the Murray-Darling Basin is far lower than the water governance structure hierarchy in China. Before the nineteenth century, the use of water resources in the Murray-Darling Bain was in the stage of anarchy. Water Councils were not set up until at the beginning of the twentieth century, when water allocation agreements were signed and governance structure began to move from agreement model to consultation model. A whole basin council appeared only in the 1980s, when the governance structure began to evolve from consultation to coordination. The process of the change in the water governance structure in the Murray-Darling Basin is closely associated with the state governance structure of Australia as a state of federation and also a typical evolutionary process in the river governance in the Western society.

Although Challen has built the model on the Western society, the model itself is generally applicable and it is a generic description of the bureaucratic structure in natural resources management, with hierarchy as a special case. Since Challen's model covers hierarchy discussed in this book, the model can be applied in the study of water problem in China. This book has generalized Challen's model as a tool for studying the property rights to water resources. This book has also localized the model and made the terminology applicable to the cases in China. After these modifications, this book terms the model as water rights hierarchy conceptual model, used specially for the study of China's water management.

3.4 Hierarchy Model Descriptive of China's Water Rights Structure

Basing on Challen's model, in this book a 'water right hierarchy conceptual model' is developed. It has two components: decision-making entities and allocation mechanism. Allocation mechanism also includes the entitlement system, initial allocation mechanism and re-allocation system. That means the water rights structure include decision-making entities, entitlement system, initial allocation mechanism and re-allocation mechanism.

3.4.1 Decision-Making Entities

Decision-making entities are de facto policy makers who hold the property rights to one or multiple rights to water resources, including the right to allocation, right to withdraw and the right to use. Water rights holders may be natural persons or corporate persons, including governments at all levels, government institutions, social groups, autonomous organizations and enterprises. In practice, all the entities at the multiple levels from the state down to individuals have the power to control the use of water resources. In contemporary China, decision-making entities may be abstractly divided into four categories according to their nature at different levels:

(1) central (basin) decision-making entities, which are the water rights holders at the state level. They include the State Council and departments in charge of water resources and the management organizations in the seven major river basins commissioned by the State Council or administrative departments in charge of water resources; (2) local decision-making entities, which are water rights holders at the regional level, including local people's governments (province, prefecture and county) and their administrative departments in charge; (3) group decision-making entities, which are water rights holders at the group level, including irrigation control organizations at all levels, water supply organizations or water supply enterprises; (4) users, which are the end level water rights holders, including enterprises, government institutions, farm households, families and individuals that use water. This book considers water withdrawal units subject to water license management as group decision-making entities, despite the fact that some water withdrawal units are end users, because water withdrawal behavior is usually collective action and in most cases is implemented by water supply organizations.[1]

As all the decision-making entities at the same level have the same type of objectives and decisions, the nature of water rights they hold are the same. Decision-making entities at different levels have different types of objectives and decisions, they are therefore different in the nature of the rights they hold. The first objective pursued by the government is to ensure water security within the administrative regions and the policies are to allocate rationally the water resources among different regions, different industries, different groups and between production, life and ecosystems. So, governments, central or local, all hold the rights to allocate water resources. The nature of water rights possessed by groups is the right to allocate resources within the scope of water withdrawn. The nature of water rights held by users is the right to use water. The rights to allocate water resources held by the central and local governments and water withdrawing groups are property rights in nature, with differences only in the targets and amount managed. The differences in the rights of decision-making entities at the same level are mainly differences in the scope and number of objects of rights.

The nature of water rights held by decision-making entities at the same level is the same because it is determined by the process of using water resources. Enterprises, farm households, families and individuals are always the end consumers of water. No matter how the water rights are allocated, they are always the ones that exercise the use right. The collective action in the development and utilization of water resources determines that water supply is an organized group behavior. Water using groups need to acquire the rights to withdraw water and at the same time allocate such rights within the groups. They therefore need a certain allocation power. Government organizations exercise the function of public management. As

---

[1]The term "water withdrawal", according to the Water Law of China, refers to acts that withdraw water from rivers, lakes or underground. Although some water withdrawal units are not end users, this book regards all the holders of water licenses as group policy-making entities. Chapter 6 will go into detail, which reveals the rationality in the simplification.

water resources have a strong nature of public good, the government does not only exercise public management functions in the general sense, but also exercise directly property management functions. So the government needs to hold macro allocation rights.

3.4.2 Water Rights Allocation Mechanism

When water resources are scarce, decision-making entities at the same level need to divide up the rights. When the scope of scarcity of water resources expands, first to the user level and then to the regional level and ultimately to the whole basin, all the decision-making entities have to demarcate their water rights. According to the four levels mentioned above, water rights allocation has to be done at least at three levels: the state to local, local to group, and group to user. The nature of rights allocated at the three levels is different. The rights allocated at the central-to-local level may be regarded as rights to allocate resources; the water rights allocated at the local to group level is the right to withdraw water; and the water rights allocated at the community to user level is the right to end use of water.

At whatever level, the rights among decision-making entities at the same level must be properly demarcated, requiring a complete right allocation system. The allocation of water rights among decision-making entities at the same level includes entitlement system, initial allocation mechanism and re-allocation mechanism (Challen, 2000). First is the entitlement system. An entitlement system is defined as a mechanism for physically dividing a resource between potential property right holders. The easiest conceivable way is allocation based on the physical quantity, that is, defining the quantity of resources due to different decision-making entities. Such direct allocation of water quantity is resource quota. Other indirect methods to control the quantity of resources allocated by decision-making entities may also be used, such as limiting the number of projects for withdrawing water, limiting the depth of water extraction wells, banning high water consumption crops and limiting areas of farmland for irrigation. Although these methods do not provide the quantity of water to be allocated, they can reach the objective of dividing up the water quantity all the same. Such indirect method is input quotas.

Second, initial allocation mechanism. This is an allocation method amongst decision-making entities at the same level. It is divided into two categories: administration-based and market-based methods. The administration-based method is to allocate water rights amongst decision-making entities at lower levels by resource administrators by virtue of administrative decisions. Market-based method is to allocate water rights amongst competing decision-making entities at lower levels by employing the price mechanism, which may be designed to operate in multiple ways, such as auction, leasing, joint stock cooperation, and sharing of investments. When there is no official initial allocation mechanism, water rights do not necessarily mean not allocated but possibly initially demarcated de facto, possibly among decision-making entities through non-official rules to self-demarcate and exercise the rights or possibly by relying on natural forces, such as first occupancy principle or the principle of giving priority to those who have the strength to withdraw water. Natural force can also be regarded as a rule of allocation. If the result is unacceptable, the official allocation rules will be introduced, such as administrative means or market means to change the pattern of initial allocation by natural forces.

Third, re-allocation system. With the changes in economic and social conditions, there might be changes in water use by all decision-making entities, thus necessitating the adjustment of water rights initially allocated. Like the initial allocation mechanism, the re-allocation mechanism can also be divided into administrative and market. By market mechanism, it means the transfer of water rights among decision-making entities at the same level through price mechanism while, by administrative mechanism, it means to adjust water rights among the competing decision-making entities at lower levels by taking administrative decisions. The adjustments of initially allocated water rights may be short-or long-termed. By short-term adjustment, it means to transfer the water use quotas of the upper reaches to lower reaches on the temporary basis. If the water use quotas are transferred for use by lower reaches infinitely, it is a long-term adjustment.

Such dichotomous classification of administrative means and market in the entitlement system, initial allocation system and re-allocation system and the two methods of resource quota and input quota in the entitlement system are simplified only in theory. In reality, the pure market and pure administrative means in the allocation of water are two extremes. The institutional arrangements in reality are often the combination of the two. The same is true with the dichotomous classification of resource quota and input quota The allocation of water quantity in reality is often the combination of the two means.

If the water rights are to be clearly defined among decision-making entities at the same level, it requires a complete water rights institutional system. The entitlement system, initial allocation mechanism and re-allocation mechanism are part of the system. The complete water rights system is classified into three categories: (1) water rights definition system, which is the rules of allocation in all circumstances; (2) water rights enforcement system, which is a system that can ensure the implementation of the allocation rules in practice; and (3) water rights sustentation system, which includes other systems that indirectly ensure the implementation of the last two categories of systems. The classification is shown in Table 3.4. The three categories of systems may be subdivided into a number of other mechanisms. Table 3.4 shows a number of examples, which do not distinguish the allocation systems at different levels.

The expanded classification method as shown in Table 3.4, the entitlement system, the initial allocation mechanism and re-allocation mechanism only cover the water right allocation system. In practice, even if the three allocation rules are complete, their efficiency depends on the exercise and safeguard system for ensuring their implementation. Challen only studies the entitlement system, initial allocation mechanism and re-allocation mechanism, probably because he assumes that there is no problem with the implementation of these systems. This book follows

Table 3.4 Classification of water rights institutional system

Institution	Category	Explanation
Allocation	Initial allocation mechanism	Initial allocation of water rights, including division of rights
	Re-allocation mechanism	Ways of adjusting or transferring water rights
	Temporary adjustment mechanism	Water volume adjustments under special circumstances
Enforcement	Monitor & control mechanism	All kinds of constraint means for overseeing the execution of allocation plans
	Incentives mechanism	Means for using water according to allocation contracts, including interests compensation
	Punishment mechanism	Means for punishing acts that have violated the allocation contracts
Sustentation	Information supply mechanism	Hydrological testing, water use monitoring, water volume statistics and information disclosure
	Interest integration mechanism	Interests expression, conflict resolution, democratic consultation and common understanding
	Assurance mechanism	Abilities of normal operations of institutions, with research projects and funds put in place

the methodology developed by Challen in the theoretical parts, that is, the system that defines the water rights among decision-making entities includes entitlement system, initial allocation mechanism and re-allocation mechanism, which are termed as allocation mechanism. The expanded classification method shown in Table 3.4 will be applied in the part of empirical study.

3.4.3 Water Rights Hierarchy Conceptual Model

With the aggravation of water shortages, the scope of water rights allocation is also expanded. When water shortages have expanded to the inter-provincial basins, the water rights allocation system will become hierarchical from the state down to users. This multi-tiered system is made up of two major factors: decision-making entities (water rights holders) at all levels and the water rights allocation mechanisms between levels. Decision-making entities have been simplified to spread in the four levels state, locality, group and user. The allocation mechanisms among various levels include entitlement system, initial allocation mechanism and re-allocation mechanism. In general, water rights are held vertically by decision-making entities at all levels and at the same level they are held by decision-making entities. The water rights structure has thus assumed a multi-tiered framework as shown in Fig. 3.1.

The nature of water rights held by decision-making entities at all levels is different. When water is allocated in the whole river basin, the water rights structure

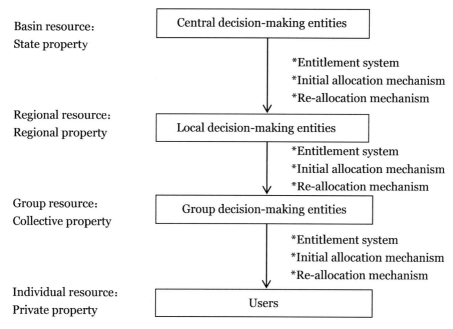

Fig. 3.1 Water rights hierarchy conceptual model

from top down as shown in Fig. 3.1 is: state-owned property rights held by central decision-making entities, which is in fact the water allocation power for the whole river basin; property rights held by decision-making entities at the local level, which is the water allocation power in authorized areas; the collective property rights held by water withdrawing groups, including water allocation powers with regard to the right of withdrawal and scope of water supply; and private property rights of end users, which is the right to use water. The higher level decision-making entities have water allocation powers over the decision-making entities at a lower level.

The object of property rights held by decision-making entities at a level is the resource shared by decision-making entities at the next lower level. The basin resources are shared by local decision-making entities while regional resources are shared by group decision-making entities. Group resource is shared by users. The common resource of decision-making entities at a given level is divided by the corresponding allocation mechanism. When the water allocation power has not been extended to the whole basin, basin-wide or regional resources are in the state of open access. Only when a perfect allocation mechanism is set up, the basin-wide or regional resources are de facto property rights owned by the state or region. When the allocation mechanism is not perfect, the de facto property rights are something between open access and state/regional property rights.

The water rights hierarchy conceptual model developed in this book is similar in form to the institutional hierarchy conceptual model advanced by Challen. But this book has modified Challen's in the following aspects: (1) Challen's model is a generic model for the management of natural resources at different levels. The model in this book is a specific application model in the management of water resources and is therefore defined as water rights hierarchy model; (2) The model in this book is the specific application of Challen's model in China and is therefore divided into four levels, namely, central, local, group and user; (3) The model in this book is a description of the hierarchical water rights. All the decision-making entities in the models used in this book are indicated in full line text box to reflect the roles of all decision-making entities in the water rights allocation. This is a specific reflection of the characteristics of the hierarchy structure while in Challen's model, the decision-making entities at a higher level are indicated in dotted line text box if there is no allocation mechanism at a given level, meaning that the water rights allocation does not have any role to play.

3.5   Characteristics of Hierarchical Water Rights Structure

3.5.1   Comparison of Water Governance Structure in China and the West

Challen studies river governance of the Western countries and his theories are built on the European and American civilizations. China is entirely different from the European and American civilizations in the logic of origin and also in the social structures. We call Chinese civilization a hydraulic society while the European civilization a contract society. In a hydraulic society, the residual control power is held by the upper most decision-making entity, with the power delegated from top down. In a contract society, the upper-most decision-making entity (federal government) are based on agreements signed with various states while the federal government is based on civil autonomy. The decision-making power of federal and state governments is endowed by written Constitution, subject to restrictions by decision-making entities at lower levels, which hold the residual control power (civil society), with the power phasing out from down up.

Challen's property hierarchy model is a generic description of water rights structure, with focal points revealing the characteristics of each level of the water rights structure, and different control power being held by different decision-making entities. Although the model is extensively applicable in describing the water rights structure in different basins of different countries, but it cannot cover up the essential disparities in the civilization background. For instance, when water scarcity is limited, there is only water rights allocation at the group/user levels and this is the same with both ancient China and Australia before the end of the nineteenth century.[2] If viewed from the form, the two are all allocation mechanisms at the group/user levels and they are the same in form. However, it would be utterly wrong to presume that Ancient China and Australia before the end of the nineteenth century had the same water rights structure.

Both China's water governance structure and state governance structure are hierarchical, with the power delegating from up down. In such a social context, even there are no allocation mechanisms at levels above group, the decision-making entities still display their important roles in the allocation of water rights, including laws and policies for the management of water supply, the building of water resource projects, water affairs personnel management and settlement of water disputes. But in a Western society like Australia, there is the tradition of civil society management, with the power delegating from down up. In such social context, if there is no allocation mechanism at the above-group levels, the decision-making entities above the group level would have little to play their roles and the water rights management at the group/user level was highly autonomous.

We have used Fig. 3.1 to characterize the hierarchy governance structure. This is an institutional hierarchy system in its real sense or we may say that this is a set of nested rules. This chart is, in fact, an expansion on the basis of water rights hierarchy model, which generalizes the allocation mechanisms into a system between decision-making entities at two levels, including the three level system of central, local and group, with the formulation and implementation of the rules supplied by the decision-making entities at the corresponding levels. In this hierarchy system, the institutions at the upper levels have the powers of restricting the institutions and decision-making entities at the lower levels and the decision-making entities at lower levels must subordinate to decision-making entities at the upper levels. Policies on the use of resources are made under the dual restrictions by institutions and decision-making entities at the upper levels.

Figure 3.3 shows the reverse form of the hierarchical structure, which may be regarded as a typical governance structure in the European civilization. Though similar in form, Figs. 3.2 and 3.3 are different in the following two aspects. First, in Fig. 3.2, the central decision-making entity in the upper-most end is the residual control right holder while such residual control right holder appears in the bottom in Fig. 3.3. Second, Fig. 3.3 adds institutions at the Constitutional level. The differences have determined that the two power structures are entirely different. In the system shown in Fig. 3.2, the central decision-making entity holds the unrestrained powers and can exercise unlimited intervention in the acts of decision-making entities at all levels below it. But things are just the reverse in the system shown in Fig. 3.3. Decision-making entities hold the decision-making rights endowed by the established system and exercise the power within the framework of the established system, with the residual control rights exercised by users at the

---

[2]See Ray Challen, Institutions, Transaction Costs and Environmental Policy, pp.116–119.

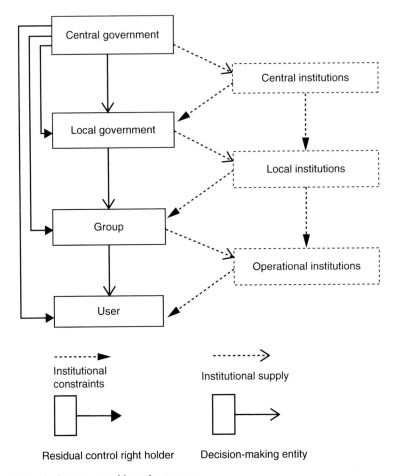

Fig. 3.2 Typical governance hierarchy structure

bottom level and the decision-making entities at the lower levels are the constraining force for the acts of decision-making entities at the upper levels.

The difference in the water governance structures in China and the West is mainly the result of the differences in the state governance structure. Challen's property right hierarchy model cannot mirror such differences.[3] This is the defect

---

[3]In the forms of the models, I proposes the following methods to reflect such differences: in the hierarchical structure, all the policy-making entities are written in solid box. If it is not a hierarchical structure, only the policy-making entities under the allocation mechanism can be written in solid box and those above are expressed in dotted text box. In Challen's works, he used dotted text box to characterize the policy-making entities in the Murray-Darling Basin in the early period. In the water rights structure in the Yellow River basin in later chapters, all the policy-making entities will be shown in solid text box, even in the structure during the Qin and Han periods.

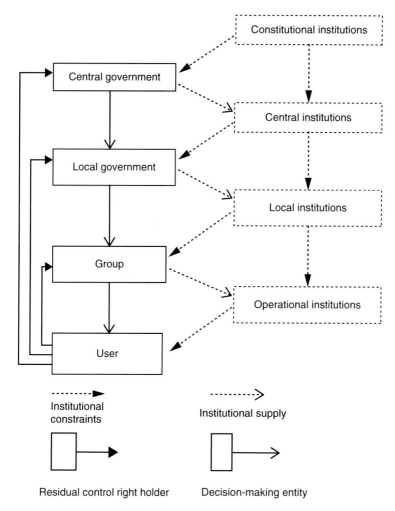

Fig. 3.3 Reverse of governance hierarchy governance structure

existing in the method to study the water right structures with different cultural backgrounds. But the model remains an effective method in the cross-basin and cross-time space comparative studies of similar cultural backgrounds. Challen uses this method to study the evolution of water rights structure over the past more than 200 years in the Murray-Darling Basin of Australia. The method is used in the following chapters and sections to study the evolution of the water rights structure over the past more than 2000 years in the Yellow River basin of China. But it is inadvisable to make a simple horizontal comparison of the two countries in water rights structure only from the descriptions provided by the model, although the two models are very similar in form. So, before taking up study of water rights structure, it is necessary to devote a large space to discussing the necessity of water governance structure. Water governance structure provides the basis for understanding water rights structure and is also the general pre-requisite for using the 'water rights hierarchy model'.

3.5.2 Characteristics of Property Rights in a Hydraulic Society

Marx noticed very early the special features of property rights in the east when he comments that the non-existence of real private ownership of land is a key to understanding the whole orient (Binder, Rhodes, Rockman, & Jessop, 2008). In *"Oriental Despotism: A Comparative Study of Total Power"* (Wittfogel, 1957), it is stressed that in the East private property has always been in a weak position. Property holders are unable to restrict the powers of the state by organizing their forces on the basis of property. That means private property in a hydraulic society is not 'power' property, but 'beneficial' property. It is argued that the private property right in ancient China is generally weak, which means that the property rights of government are extremely strong and the exercise of the power without restraints, covering all areas and having the ultimate ruling on private property. Such power is exhibited as powerful property right but it is weak in terms of the beneficial property (Liu, Wang, & Chen, 2014b; Perdue, 1990).

A review of the history of law in ancient China, no traces of individualism was found. There were only ideas that placed collective interests above private interests. The law of ancient China makes no distinction civil law and criminal law, and individuals are liable to others and society, without giving particular attention to private rights. The ownership concept in ancient China is relative. So long as there is the need for collective interests, the exercise of property rights is restricted. But property right economics in the modern Western world, complete property rights mean a group of rights or a right system to a given right. In most cases, property rights are incomplete or attenuated due to restrictions by society. According to this view, the property rights held by individuals in a hydraulic society is highly incomplete and they are seriously attenuated and highly defective due to social restrictions.

Further, we divide property rights in a hydraulic society into two parts, which is, in fact, the relative division of the bundle of rights. Part of them enters the public areas to become public rights to be exercised by the government and another part goes to the private area to be used beneficially by individuals. Public rights have absolute superiority over private interests. On the one hand, a hydraulic society lacks the law to ensure private property. On the other hand, there are no forces in the society to rival the state public power. This form of ownership is public ownership, which is known as imperial ownership in ancient times and state ownership in modern times. The public ownership is the basis for the government to exercise public power while individuals only have the right to use or manage the property.

The public power in a hydraulic society has lawful sources at the beginning from the protection of private interests and theoretically it should protect the rules for ensuring public interests. However, once the unrestricted public power has become a fact, due to the coercive power over the private use right or management right of individual unconditionally, public power has often become a force to plunder private interests.

The characteristics of proper rights in a hydraulic society run through all the property right relations of all physical assets, especially so with water resources. The private ownership of upper stream water had never been mentioned in any water law in any period of ancient China. Nor were there any rules about the inheritance of land or the ownership of water under the land. This is because of the special non-exclusive nature of water, which had always been regarded as public property. Water was developed and allocated by public power and individuals benefit from the use of water without impinging upon the public interests and on the contrary they must subordinate to public interests and undertake corresponding obligations. In the term of modern property right economics, individuals enjoy only extremely imperfect property rights.

3.5.3   Characteristics of Water Rights of Hierarchical Structure

The characteristics of a hydraulic society is summed up into the following six aspects, which are applicable to water rights both in broad and narrow senses.

First, it focuses on public ownership of water rights. In ancient China, water was publicly owned. "Mountains, rivers, lakes and wetlands were not owned by private persons. No private ownership was allowed. This is an established system" (Fu, 1988, p. 60). Public ownership of water rights means that all water resources are allocated and used directly by the government that has the decision making power. Almost all important irrigation systems were built and maintained under the direct control of the government. All canals, dykes and dams, no matter who built them, were subject to the control by the government, and only small irrigation works without large water allocation systems were left to the management by private persons. The public ownership of water rights in ancient China is, in essence, the ownership of the imperial court. The supreme rulers held the supreme power of the society and had the ultimate power over mountains, forests, pools and wetland and other natural resources, that is, it held all the residual control rights over water and other natural resources. The public ownership of water was turned into state ownership in modern society.

Second, water rights are exercised by the bureaucratic hierarchy. In ancient China, water rights were of public ownership in name only. It was, in fact, owned by the imperial court. In practice, the rights were exercised by the bureaucratic system that represented the imperial court, including water resources commissioners and local officials in the general sense. In modern society, imperial power has been evolved into state power exercised by the central government and water rights are still exercised by the bureaucratic system. This shows that in the water rights structure of China, the central government holds the residual control rights and is the ultimate principal of rights. Localities and groups are the agents of the central government. Theoretically, the decision making power of the official system is granted by law, that is, the series of laws, orders issued by the central government, together with the corresponding powers of officials and how they would exercise them. In the process of exercising the water rights, officials would have a certain degree of freedom in taking decisions, that is, having a certain degree of residual control right. The residual control power held by the bureaucratic institutions and the central government is the same in nature, which may become power to benefit the society or power to plunder the private interests in the name of the public.

Third, the extreme limitation of private rights. The private rights in a hydraulic society are greatly attenuated, especially in the development and utilization of water resources. In the pattern of public ownership and domination by bureaucracy, private interests were entirely subject to the manipulation by public power and the private rights and benefits associated with water were very attenuated, legally or in practice. Just as what is mentioned above, private ownership of water had never been mentioned in the water laws of any period in ancient China. Water had long been regarded as public property in China and therefore must be developed and allocated by the public power. Private persons can only benefit from the use of water without impinging upon the public ownership. They must subordinate to the public interests and undertake corresponding obligations. In principle, public ownership and private interests of the public are compatible, with private interests nested in collective or state interests, although it is not always so in practice.

Fourth, superior levels have absolute superiority over lower levels. The unique natural geographical characteristics of China determine the highly hydrological uncertainties in the use of water resources since ancient times. In an environment with a high degree of uncertainty, the option for a hierarchical governance structure can ensure security to the maximum. In this structure, authorities at higher levels have superiority over those at lower levels in decision-making and can get involved or intervene in any policies taken by authorities at lower levels. The central decision-making entity has the highest priority in intervention. Consequently, the users at the bottom level are always the first to undertake obligations and the last to enjoy benefits. This allocation of decision-making power between higher and lower levels in the hierarchical governance structure determines that, apart from the supreme level, no decision-making entity has any definite rights. The lower the levels are, the greater the uncertainties in the rights held. The rights held by users at the bottom level have the greatest uncertainties, likely to be deprived of altogether at any time.

Fifth, uncertainties of rights among decision-making entities are at the same level. The hierarchical governance structure has a clear definition of the rights and obligations between higher and lower levels, which is meant order and obey-order relationship. But the relations among the decision-making entities at the same level are uncertain. The information flow in this structure mainly from down up or from up down, lacking the horizontal movement. Such information flow leads to a lack of communications among decision-making entities at the same level. There are more conflicts than cooperation among them. Their cooperation is mainly coordinated by authorities at a higher level. In practice, it is expressed in the tradition of lacking horizontal cooperation. So, apart from the de facto right held by decision-making entities owing to natural factors, the rights held by decision-making entities are mainly decided by the decision-making entities at a higher level. The relative priority with regard to water rights among decision-making entities at the same level is, to a large extent, decided according to the preferences of the decision-making entities at a higher level.

Sixth, state-contingent of rights is significant. All the above determine that water rights are de facto contingent to a given state. For a given entity, the decision making right it holds has the characteristics of state contingent. Whether or not it should hold the right and to what extent it should exercise the right are contingent to whether or not the entities at higher levels get involved, to what extent it gets involved and the demand for rights of the decision-making entities at the same level. The rights with state contingent are originated from the uncertainties of the environment, including natural environment and social environment, especially in the allocation and use of water resources. That determines that water rights have a significant state contingent.

The above characteristics of water rights are determined endogenously in the hierarchy governance structure. The choice model of governance structure in Chapter II shows that, in order to maintain stability of the hierarchy structure, it is necessary to lower management costs, which leads to corresponding arrangements of control power (decision-making power). The central government holds theoretically all the residual control power, but in practice, it delegates a certain residual control power to local and group decision-making entities. The residual control power in the hierarchy structure is attenuated from up down, with the users at the bottom level holding little decision making power. That has made it inevitable for the water rights to exhibit the above six characteristics. The main features of water rights in China, no matter the attenuated private rights or priority of higher levels over lower levels or state-contingent, are, in essence, associated with the hierarchy governance structure and can be rationally explained from the angle of lowering management costs within the hierarchy structure.

Chapter 4
Economics Theory of Water Rights Structure

This chapter provides an economic explanation to the water rights structure. First of all, it reviews some literatures on property rights in the West, with emphasis on the thought and methodology of the modern Western property rights studies, especially the tide of economic explanation. Then it discusses about the static and dynamic theories based on the existing theoretical studies by Challen (2000), in order to explain the reasons why the water rights structure is hierarchical, what is the intrinsic logic for the option of water rights allocation regime and what is the law governing the dynamic changes of water rights structure. The chapter focuses on the discussions on the motivation mechanism for introducing market elements into the hierarchy system, which is unique in China and also something that has not been touched in Challen's theories. For this purpose, this book introduces the intermediate variable of the 'quality of water rights' to expound on the mutual exclusiveness between administrative and market allocation of water rights. On this basis, the book points out that the hierarchy structure is, in essence, an administrative control system naturally incompatible with market mechanism. However, with the changes in the external environments, especially the growing severity of water resource scarcity and the rise in the internal management costs or the lowering of cooperation costs, the hierarchical structure would develop the motivation for introducing market mechanism, a process concomitant with the lowering of the degree of hierarchy. In fact, market mechanism has a very limited role to play in reality, as its introduction and effective operation require stringent conditions, including lower enough operational and institutional change costs.

© Springer Nature Singapore Pte Ltd. 2018
Y. Wang, *Assessing Water Rights in China*, Water Resources Development and Management, DOI 10.1007/978-981-10-5083-1_4

4.1 Modern Property Rights Theory in the West

4.1.1 Original Theory of Property Rights

The institutional theory in which the early property rights were originated attempted to explain the option for exclusive private property rights or the opposite common property rights and identify the forces that make property right structure get closer or far away from exclusive private property rights. This was later termed as 'original theory of property rights' (Thráinn Eggertsson, 1997), which explains why exclusive rights were established or not established by taking into consideration only the costs and benefits under the action of economic factors.

The article *"Toward a Theory of Property Rights"* (Demsetz, 1967) is an early classical reference for the original theory of property rights. For the first time, the paper advances the idea that property rights develop to internalize externalities when the gains of internalization become larger than the cost of internalization. Increased internalization, in the main, results from changes in economic values, changes which stem from the development of new technology and the opening of new markets, changes to which old property rights are poorly attuned. Demsetz uses his theory to explain the introduction of private ownership of land among Indian hunters in the eastern part of Canada in the early years of the eighteenth century. He points out that before the fur trade became established, the Indians hunted beavers only for their own consumption of meat and furs, exclusive rights were nonexistent. The externality was clearly present, but there did not exist anything resembling private ownership in land. With the development of commercial fur trade, an increase in demand led to a sharp increase in hunting and the Indian even marked off their own hunting ground to make the rights of land exclusive, ultimately leading to the segmentation of the land. Analysis is furthered by the argument that the Indians in the American Southwest failed to develop similar property rights because of the relatively high costs and low benefits from establishing private hunting lands, as in the American Southwest, there were no animals of comparable commercial importance to beavers, and the animals of the plains were mostly grazing species who wandered over wide tracts of land (Demsetz, 1967, 1988).

Anderson and Hill (1975) present a graphic model involving a marginal cost function and a marginal benefit function for investment in the definition and enforcement of property rights and identify critical shift parameters for each function. Their model indicates that a fall in the price of exclusion inputs or a change in exclusion technologies shifts down the marginal costs function and, ceteris paribus, increases exclusion activity. Anderson and Hill apply this model to evolution of exclusive rights of the utilization of land, water and cattle on the Great Plains of the American West during the second half of the nineteenth century. North and Thomas (1973) use the original theory of property rights to provide a new explanation for the development of agriculture in human prehistory. Their model indicates that while plants and animals were relatively abundant, the costs of establishing exclusive rights to these resources exceeded the potential gains, and natural resources were used as common property. As the human population increased relative to the constant resource base, and competition among bands stiffened, open access led to diminishing returns in hunting. At the margin, settled agriculture gradually became more attractive than hunting, although agriculture required the costly establishment and enforcement of exclusive rights. The shift of man's major economic activity from hunting and gathering to settled agriculture created an incentive change for mankind of fundamental proportions. The incentive change stems from the different property rights under the two systems. When common property rights over resources exist, there is little incentive for the acquisition of superior technology and learning. In contrast, exclusive property rights which reward the owners provide a direct incentive to improve efficiency and productivity, or, in more fundamental terms, to acquire more knowledge and new techniques. It is this change in incentive that explains the rapid progress made by mankind in the last 10,000 years in contrast to the slow development during the million-year long era as a primitive hunter/gatherer (North & Thomas, 1973).

The original theory of property rights has been brought into depth by scholars. When resources are more valuable, the rules for defining property rights have become clearer (Murray, 2016). Umbeck (1981) studies the development of mineral laws, tracing their development of gold mine in Nevada State of the United States in the nineteenth century. He has proved that in the early years, while the rights structure was still incomplete, the heated competition for land following new ore discoveries increased the miners' demand for more exclusivity, which in turn led to increased specificity of the mining law. In terms of internal governance costs and the costs of exclusion, when the internal governance costs drop, equilibrium is reestablished by associating larger commons; when the costs of exclusion drop, the equilibrium is established by associating smaller commons (Emel & Brooks, 1988).

Property rights is also explained by 'rent dissipation' theory, and common property rights would bring about rent dispersion, which may economize the cost of exclusiveness. Although private property rights may avoid rent dissipation, it has to pay the cost of establishing exclusiveness. An asset has become commonly owned in that the rent it obtains is lower than the cost of enforcing exclusive rights. (Cheung, 2000). Barzel (1989) advances the property rights methodology in the 'public domain'. As transferring, obtaining and protecting property rights need to pay transaction costs, any right is not completely defined. The rights that are not defined have left part of the valuable resources in the 'public domain' for people to capture. The value of all the resources in the public domain is termed as 'rent'. For every potential rent seeker, the marginal cost is equal to the marginal increment of the rents it has obtained under the rights enjoyed (Barzel, 1997, 2000). The property right paradigms of Cheung and Barzel are in fact a new way of expressing the original theory of property rights.

The original theory of property rights reveals that exclusive rights to an asset are established and enforced when potential owners expect positive net gains from exclusivity. With rising marginal costs of enforcement and falling marginal benefits, exclusive rights are seldom complete. Furthermore, optimizing owners seek enforcement at margins where costs of measurement and enforcement are low, which means that property rights are not completely defined and the property rights of an asset in the real world is at an 'equilibrium' or 'optimal' level of definition. The 'equilibrium' definition level of property rights is a dynamic process. As the raising of exclusiveness of property rights has to pay costs, only when the marginal gains are bigger than the marginal costs in defining the property rights, motivation would be present to formulate rules to define property rights until the marginal costs are equal to marginal benefits in defining the property rights to attain a temporal institutional equilibrium. When the benefits of defining the property rights rise or the costs of defining property rights drops, the static institutional equilibrium would be broken and re-definition would start until a new institutional equilibrium is achieved.

4.1.2 Expansion of the Original Theory of Property Rights

The original theory of property rights holds that the decisions made about property rights are only associated with the private costs and gains, without touching the free rider problem and other factors affecting group policy making or the political process. When political process plays an important role in defining property rights, the property rights theory without the involvement of government is not complete and instances of government's role in promoting or obstructing institutional arrangements for economic growth exist extensively. In the book *"The Rise of the Western World"*, North and Thomas (1973) comments that governments take over the protection and enforcement of property rights because they can do it at a lower cost than private volunteer groups. However, the fiscal needs of government may induce the protection of certain property rights which hinder rather than promote growth; therefore we have no guarantee that productive institutional arrangements will emerge.

According to the original theory of property rights, the existence of low-efficient property rights structure in reality must be due to some transaction-cost constraint. Just as it is revealed by the Coase Theorem (Gjerdingen, 2014), with the positive constraint of transaction costs, different property rights structures would lead to different economic results; when the transaction costs are high, the government may allocate property rights directly to individuals or use other methods to redefine property rights so as to maximize wealth. But in reality, just as is pointed out by North, the government will not necessarily define efficient property rights structure to maximize the net social wealth, due to the interaction of interest groups.

The adjustment of property rights structure would make the benefits of interest groups rather uneven. If the groups of immediate interests have a strong influence on the government, the establishment of efficient property rights structure would be obstructed. Libecap (1986) describes interest groups in the property rights theory as competition forces tend to erode institutions that no longer support economic growth, and predictions regarding the way in which property rights arrangements respond over time to changing economic opportunities. Distributional conflicts arise when property rights are coercively redistributed by the state with little or no compensation. Disadvantaged parties will oppose the new arrangement, even though it allows for an aggregate expansion in production and wealth. Accordingly, analysis of the likely winners and losers of economic and institutional change and their interaction in the political arena in specific settings is necessary before the observed pattern of property rights can be understood (Thrainn Eggertsson, 1990).

Lin (1994) makes distinctions between enforced institutional change and induced institutional change and explains different paths for supplying efficient institutions. Induced institutional change refers to spontaneous change in response to the opportunities induced by institutional disequilibrium, including the change in the institutional option combinations, technical conditions, institutional demand or other institutional arrangements. The free rider problem is concomitant of collective actions due to the fact that institutional change may result in inadequate institution supply, thus making it necessary for law-enforced institutional change (J. Y. Lin, 2013). However, due to the existence of interest groups, enforced institutional change cannot ensure adequate supply of efficient institutions.

Whether or not the government can enforce a property right structure depends on what degree the government can overcome the influence of groups' immediate interests. It is illuminated that special-interest groups require long periods of social tranquility to overcome free riding and to organize as pressure groups. Once pressure groups are organized, they are likely to seek various privileges that can strangle economic growth. But, on the other hand, turbulent periods tend to uproot pressure groups and make rapid economic growth possible (Olson, 1982). In other words, when pressure groups are weak, the state provides a structure of property rights that is consistent with the original theory of property rights.

4.1.3   Efficiency Standards of Property Right Structure

Mainstream economists since Adam Smith all hold that different property right structures lead to different economic output and the exclusive private property right corresponding to market mechanism is regarded as the most efficient property right structure. However, exclusive property rights extensively exist only in the ideal world in which transaction costs are zero. In the real world in which transaction costs are positive, pure private property right structure is rare and extensively existing is a diversity of property right structures. But are all the non-exclusive property right structures efficient in using resources in the real world? On this issue, neoclassic economics and new institutional economics are divided.

Neoclassical economists criticize property right structures in the real world as inefficient. They hold that the imperfect property right structures in the realities lead to all kinds of externalities, causing market failure and restricting the role of market mechanism. When market failure occurs, neoclassic economics advocates for direct intervention by the government. As the government is facing more serious failures, they hold that the government should use as much as possible such economic measures as defining property rights, levying taxes, providing subsidies, introducing licensed trading instead of direct control that is low in efficiency to correct market failure and let market display its role to the maximum.

This analytical paradigm has drawn protests from several economists, who deem the comparison between what is real and what is ideal 'unreasonable'. Demsetz (1967) argues about a relevant choice, as between an ideal norm and an existing imperfect institutional arrangement, is crucial for analysis. This nirvana approach differs considerably from a comparative institution approach in which the relevant choice is between alternative real institutional arrangements (Thráinn Eggertsson, 1997). Cheung's (2000) views that the real world is efficient which has a big impact on the development of new institutional economics. Before Cheung, most economists believe that crop sharing tenancy and fixed rent contracting and independent planting by owners would lead to inefficient allocation of resources on the ground that in crop sharing, and part of the output of tenants is taken away as rents, thus reducing the incentives of tenants' labor and investment. Cheung points out that, after taking into account natural risks in agricultural production and different contract arrangements having different transaction costs, sharing contract, like fixed rents contract and independent planting by owners, can achieve optimal allocation of resources. The option for agricultural contracts in different areas is determined by the transaction costs of each contract and the abilities of sharing risks (Myint & Cheung, 1970).

The key in the differences between new institutional economics and neoclassical economics lies in the introduction of positive transaction costs in the neoclassical economic analytical paradigm. As positive transaction costs result in large quantities of non-private property rights, the view that criticizes common property rights for causing waste and holds that when common property rights are turned into private property rights, waste would disappear. New institutional economics holds that the all kinds of common property rights in reality must be the institutional option for rent optimization under all constraints, which is defined by Cheung as minimization of rent dissipation under constraint conditions, that is, common property rights would bring about rent dissipation due to its inability of establishing non-exclusive rights structure. However, wealth maximization would drive people to make institutional arrangements that would reduce rent dissipation to the maximum. Common property rights are unable to create exclusive private property rights, but it can economize on exclusive costs. After taking into consideration all the transaction costs, a given common property rights structure in reality is, of necessity, the result of maximization under all constraint conditions that is, therefore, efficient.

The views about efficiency held by new institutional economics may be extended to general circumstances, that is, the institutional arrangements in reality are endogenous, therefore, the equilibrium result of maximization under all constraints,including political and economic transaction costs,are efficient. Viewing from this point, the optimal property rights structure or the most efficient property rights structure would be an idealist empty talk. The correct option is to recognize the economic rationality of the real world, explain why there are so many kinds of property rights structures in reality and identify all the constraints that determine given property rights structures. Economics may seek ways to transform reality by explaining the reality. In fact, the force of economics to transform the world is impregnated in its insight into all constraints that determine the reality, but not in the sheer criticism of the gaps between reality and the ideal world. Just as Cheung points out that the most important task of economics lies in the economic explanation of the real world. A tide of economic explanation is rising in modern economic analysis (Cheung, 2000).

4.2 Static Economic Explanation of Water Rights Structure

The preceding chapter has developed the conceptual model of water rights hierarchy. This section will use economics theories to make static explanation of this model to answer why water rights assume a hierarchical structure and what is the logic for opting for a mechanism of allocating certain rights.

4.2.1 Economic Explanation of Water Rights Hierarchical Structure

Challen (2000) makes an excellent economic explanation of the institutional hierarchy on the basis of what has been achieved in the property rights economics. He holds that the option for all institutional structures follows the logic of transaction cost minimization under constraint conditions, and a group of given institutions in reality is the minimization of transaction costs for formulating policies and creating and maintaining institutions. His studies introduce the terms of static transaction costs and dynamic transaction costs. Static transaction costs are costs associated with making and implementing decisions for resource allocation under a given institutional structure, including costs of making decisions and costs (definition cost) of achieving the objectives of resources allocation and costs of creating and maintaining institutions (enforcement costs). The static analysis of transaction costs is, in fact, the rule for minimizing static transaction costs in institutional options. Dynamic transaction costs include transaction costs of institutional change, which determine path dependency of institutional change (Challen, 2000).

Why do water resource property rights assume a hierarchical structure? What Ray Challen explains is that the decisions on the use of resources are made at different levels to minimize the transaction costs, because there are multiple decisions that need to be made for use of a resource. Each type of decision is associated with particular information requirements and patterns of interests among individuals or groups in a society. Consequently, transaction costs of making decisions on resource management may be minimized by assigning powers to make particular decisions to different levels in an institutional hierarchy. The flowing character of water resource determines multiplication of decisions makings and that is why water resource property right structure is hierarchical (Challen, 2000).

But why China's water rights structure is not only a hierarchy but also a special one? According to Challen's theory, institutional structure is the product of transaction costs minimization in a given environment. This is true with China's water rights structure. This explanation is in conformity to the theory for the operation of water governance structure in Chap. 2. China is different from the Western society in terms of water governance structure in that they have different environment constraints in water governance activities. The frequent droughts and floods in China determine that China needs a highly centralized system in the early period of civilization and the hierarchy structure could economize on transaction costs to the maximum in such an environment. The Western world does not face such grave challenges by water and a relative power sharing governance structure is the product of minimization of transaction costs. The water governance hierarchy structure in China determines the hierarchy structure of water rights. Just like the logic for the option of water governance hierarchy structure, the water rights hierarchy structure is an institutional arrangement with the minimum transaction costs under all constraints.

4.2.2 Economic Explanation of Water Rights Allocation Mechanisms

In the water rights hierarchy conceptual model, we use the three most important water rights regimes put forward by Ray Challen: entitlement system, initial allocation mechanism and re-allocation mechanism. The option for these institutions, too, follows the same logic of minimization of transaction costs under environment constraints, as Challen points out. The following is an introduction to the explanation by Challen of the three water rights mechanisms (Challen, 2000).

First, regarding entitlement system, in water rights allocation, both input quota and resource quota require the society to pay costs. In terms of the entitlement system in the allocation of irrigation water, in the entitlements allocation by resource quota, the costs include costs of information gathering, costs of knowledge sorting necessary for the manager to fix the amount of quotas, costs of rules formulation for adjusting quotas when water quantity changes, costs of measuring the volume of irrigation water, and costs of creating incentives and penalty mechanisms necessary for ensuring water is used according to quota allocated; for the entitlements by input quota, the costs include costs of collecting information and

Table 4.1 Options for water rights allocation

		Cost of market means	
		Low	High
Cost of administrative means	Low	Administrative or market	Administrative
	High	Market	Administrative or market

knowledge associated with irrigation for limiting quotas and standards for limitation and monitoring of the implementation of input quotas and ensuring of its enforcement. In the allocation of irrigation water, input quota is more economical than resource quota in the costs of measuring water used and adjusting the water volume and in fixing quota and monitoring implementation. In other words, the costs of input quota are usually lower than the costs of resource quota. But resource quota has stronger exclusiveness in the division of rights but less externalities than input quota, thus it is more favorable for the intensive use of resources; input quota has weaker exclusiveness, but more externalities, thus resulting in more extensive use of resources. This demonstrates that the water use efficiency of resource quota is usually higher than input quota.

Second, in terms of initial allocation mechanism, theoretically, there are two ways of water rights allocation, which are market and administrative means. In reality, the option for water rights allocation depends on the cost comparison between the two. Table 4.1 shows that when the market mechanism requires higher costs but yields lower benefits, administrative means is preferable. When the market cost is low but benefits are high, market is more preferable. When transaction costs are zero and the market is in full competition, the efficiency is beyond doubt. But when constrained by positive transaction costs, incomplete information and information asymmetry would distort prices and the market would not necessarily display its advantage of efficiency and furthermore, thus market mechanism also requires costs. If such costs are high enough, the use of market mechanism does not pay off.

The cost of initial water rights allocation by administrative means is associated with information, that is, information required from the decision-making entities at a lower level in order to fix a rational initial allocation. But how to get the initial allocation scheme approved by other decision-making entities at the same level? That requires information exchange to reach common understanding through negotiations. Generally speaking, the more the number of decision-making entities at the lower levels, the stronger the heterogeneity and the higher the costs of the administrative means are and vice versa. The less the number of decision-making entities at high levels, the less the cost of water allocation is by administrative means. When the number of decision-making entities is very large at the bottom level, the cost of administrative means is high and there would be bigger motivation for introducing market mechanism.

What merits special mention is that water being a kind of basic resource, the biggest difference in allocation as compared with ordinary economic assets is security, fairness and social acceptability. Although market enjoys natural advantage in terms of efficiency, the costs (external transaction) of market is very high: (1) it is impossible for market mechanism alone to produce a fair allocation scheme; nor is it possible for it to take water security into account; (2) due to big differences in different areas, industries and users, it is impossible for the initial allocation of water resources to form a fully competitive market; nor is it possible to form an equilibrium prices simultaneously; (3) the establishment of a water resource auction system per se is very costly and even if it can be established, all kinds of transaction costs in reality would prevent market from reaching an equilibrium, thus making it difficult for the auction market to operate. Relative to the drawbacks of market, the government enjoys natural advantages of ensuring water security and fairness in allocation. That explains why administrative means is often adopted in the initial allocation of water resources and why market means is adopted only in a few very special cases. In the following discussions on the introduction of market mechanism into water rights allocation mechanism, it mainly refers to the introduction of market mechanism into the reallocation mechanism.

Third, like the initial allocation mechanism, the reallocation mechanism may be summed up as being based on both administrative means and market means. The economic logic of reallocation mechanism is also applicable in Table 4.1. When the costs of market are high, the administrative means is preferable and when the costs of administrative means are high, market is preferable. The transaction costs of market allocation of water rights are associated with many factors such as social system, legal system, cultural tradition and the concept of values, water resources management system, the complexity of hydrology and the operation of the water market. In a society that has long been using administrative means to regulate water rights allocation, such as China, the infrastructure for the operation of water market is usually very feeble and the market reallocation of water rights is of necessity entailing high transaction costs and the administrative means in regulating water rights is, therefore, advantageous. But things are quite the opposite in European and American market economies, where it would be much easier to use market to regulate the allocation of water rights as the transaction costs are much lower.

What needs additional explanation is that resource quota and input quota in the entitlements system, and the administrative means and market means are not mutually exclusive. In fact, they are often used together in order to economize on the general transaction costs. For instance, water withdrawing units must pay water resources fees; and farm households have to put in their labor in order to obtain the irrigation water rights. This is a kind of combination of administrative and market means in water rights allocation.

4.3 Dynamic Economic Explanation of the Change in Water Rights Structure

On the basis of studying the latest theories about institutional change, especially North's institutional change model, an analytical framework of institutional change constrained by dynamic transaction costs is developed (Challen, 2000). Dynamic transaction costs are those required in institutional change. To the extent that the costs associated with decision making for different institutional options are a function of pre-existing institutions, institutional change will be path-dependent. Challen's framework is very useful in understanding the dynamic changes in the water rights hierarchy.

According to the 'water rights hierarchy conceptual model', a water rights hierarchy is in a state of constant change, including (1) change in decision-making entities such as in the number, the ways of interconnection and the expansion from the bottom up to the upper part; (2) change in entitlement system, such as transition from input quota to resource quota; (3) change in the initial allocation mechanism, such as the implementation of water allocation scheme marking the establishment of the initial allocation system; and (4) change in the reallocation mechanism, such as the change in the regulation from administrative to market means. The changes in the above four aspects add up to the change in the water rights hierarchy as a whole.

Motivation for the water rights structure change may come from changes in the external conditions, such as the increase in the resource scarcity, change in population and technical progress and institutional environment. New environment would lead to new equilibrium structure and water rights structure would change from initial allocation to a new equilibrium structure. However, such a change is not easy to take place, because institutional adjustment has to pay dynamic transaction costs. If such costs are very high, it would make it difficult for such a change, forcing institutional change to opt for a path with the minimal dynamic transaction costs in order to lower the costs of institutional change. As the dynamic transaction costs are the function of the initial state of water rights structure, it has made institutional change strongly 'path dependent' (McCann, Colby, Easter, Kasterine, & Kuperan, 2005).

The logic of the above changes in water rights structure explains why there is inefficient institutional structure in the allocation of water resources. If dynamic transaction costs are taken into consideration, the existing institutional structure is also efficient. This logic also explains why an institution has a powerful inertia, making it difficult for changing from one habitual institution to another. In China, for instance, the prolonged administrative allocation of water resources has found increasingly difficult to adapt to the new situation in which water scarcity is increasing. The market allocation could be profitable. However, the transition from planned allocation to market allocation would need exorbitant costs, thus making the change very difficult.

In Challen's (2000) study against the background of Australia, the change in water rights structure takes place in the form of induced institutional change, which is the main form of change in the West. But the opposite is true with China because of separation of property right and political right and low degree of autonomy of the society and the monopoly of political rights by the bureaucratic system. This determines that under the social conditions of China, enforced institutional change would be the main form in the evolution of the water rights structure while induced institutional change often plays an auxiliary role. Enforced institutional change by administrative authorities lowers the dynamic transaction costs and solves the problem of free rider concomitant of the induced institutional change. But it often encounters with the difficulty in institutional compatibility. The incompatibility of the old and the new institutions would result in the rise of the operational costs. The enforced institutional change is not necessarily always efficient.

4.4   Motivation Mechanism for Introducing Market Elements into Hierarchy Structure

4.4.1   Water Rights Quality and Allocation Mechanism

In the water rights allocation mechanism, administrative and market means are mutually exclusive or mutually replaceable. The use of administrative means naturally excludes market mechanism while the use of market means would require the reduction of administrative means. In the following, we shall introduce an intermediate variable water rights quality (θ) to analyze briefly the inter-replicability between administrative and market means in the water rights allocation mechanism.

The characteristics of property rights are multi-dimensional. Scott (1989) recognizes the following six characteristics of property rights: (1) Exclusivity—the extent to which other parties can be excluded from the item or from the flow of benefits arising from the item; (2) Duration—the period over which the rights exist; (3) Transferability—the extent to which the right may be transferred between parties and the ease with which it may be transferred; (4) Divisibility—the degree to which the property right can be subdivided; (5) Quality of title—the extent to which the property right describes the item covered by the right, the related rights and duties of the rights holder and other parties, and the penalties for violation of rights or non-performance of duties; and (6) Flexibility—the extent to which the owner of the property right can alter patterns of use of the item to meet his own objectives. The six characteristics of property rights are attenuated to varying degrees due to the constraints imposed by the society. The multiple dimensions of property rights can be expressed in a chart, with the length of straight lines to represent the degree of attenuation as shown in Fig. 4.1.

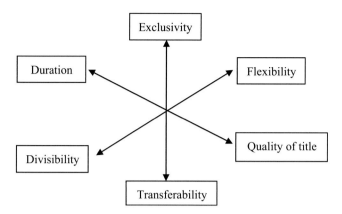

Fig. 4.1 Multiple characteristics of property rights

The characteristics of property rights mentioned above are general in nature. Usually the lower the degree of attenuation of the rights, the closer it is to pure private property rights. According to the characteristics of water resources, the book classifies the characteristics of water property rights into four dimensions: exclusivity, duration, transferability and divisibility, with the exclusivity as the main vector. Although the four characteristics are not necessarily interconnected, with the increase in the exclusivity, the degree of the other characteristics also rises. In other words, when the quality of rights improves, the 'asset nature' of water rights is also enhanced. The book defines the characteristics of water rights as the 'quality of water rights' to reflect the characteristics of water rights as a whole.

Assuming water allocation is conducted between the government and users, the multiple levels of water rights allocation are simplified into a single level analysis. The quality of water rights held by users is expressed in θ, then $\theta \in [0,1]$, $(1 - \theta)$ expresses the degree of attenuation of rights. If $\theta = 0$, it indicates the rights are wholly attenuated and the quality of water rights is extremely low and rights held are similar to things that belong to others. If $\theta = 1$, it indicates the rights are not attenuated and the water right quality is very high and the rights held are entirely private assets, freely disposable.

Now look at the relations between θ and government intervention. If θ is close to 1, it means the rights have strong exclusivity, duration and divisibility, with the water rights similar to private assets of users. Such water rights exclude government's interference and tend to realize right transfer through voluntary contracts. If the government interferes successfully, such as adjusting by administrative means and the existing rights allocation pattern, the water rights are sure to be attenuated to a large extent and the θ value is small, so that the exclusivity of rights is attenuated and it would neither last long nor is clearly defined and its restriction of transferability provides the basis for the government to adjust the rights.

This shows that θ is in close negative correlation with water rights allocation by administrative means. The larger the value of θ, the higher the cost of right

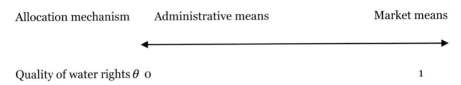

Allocation mechanism   Administrative means   Market means

Quality of water rights θ 0   1

Fig. 4.2 Mutual replacement relations between administrative and market means in water rights allocation

adjustment by administrative means; the smaller the value of θ, the lower the cost of right adjustment by administrative means. The increase in the value of θ would lead to high costs of administrative means and low costs of market means in adjusting water rights. When the quality of water rights held by users is low, there is not the prerequisite for trading among users and uncertainty of rights leads to high transaction costs and thus further lowering the possibility of trading. This is what has happened in a hydraulic society, where property rights are generally very weak, thus suppressing the role of market. With the rise in the quality or rights, the rights held by users have acquired the nature of assets and that will provide the possibility for trading. If the transaction costs are small enough, the rights turnover would take place in large quantities. That is what has happened in the western society, where high quality of property rights has provided the prerequisites for the market mechanism to display its basic role in all areas.

The above analysis unveils an empirical law governing water rights allocation, that is, high quality of property rights is corresponding to market allocation while low quality of property rights is corresponding to administrative allocation. Figure 4.2 reflects this law perceptually. Quality of water rights serves as a bridge between administrative allocation and market allocation and by observing the value of θ, it is easy to judge the option for the ways of water rights are allocated, thus bypassing the more complicated transaction cost-based analysis. What needs explanation is that water rights quality is not heterogeneous. It is endogenous in water rights structure. For instance, the property rights are weak in a hydraulic society in that its hierarchy governance structure requires as much as possible economizing of management costs. The lowering of quality of rights will provide conveniences for administrative interferences and that is an effective way for lowering management costs. So the weakness of property rights in the oriental hydraulic societies is endogenous in the social structures.

4.4.2 Hierarchy Structure and Administrative Allocation System

Water governance hierarchy is in essence an administrative control system, just like a firm. In the theory of the firm, the reason for the existence of a firm is that plan and instruction replace price mechanism. The option for water governance structure advanced by this book is similar to it. It is to use administrative control to replace political negotiations. The output of water governance is the collective security and that is a political process for more of its supply. This is different from a firm. The water governance hierarchy is, therefore, to use enforced administrative coordination to replace political transaction but not market transaction. Whatever the case, hierarchical structure economizes on the bargaining costs. In the theory of the firm, the costs economized by a firm are the transaction costs of what Coase (1960) means. Such costs economized by the water governance hierarchy are defined as cooperation costs in this book. Water governance hierarchy, in the general sense of water affairs, is naturally applicable to the allocation and use of water resources. Water rights structure is the specific application of water governance structure in the allocation and use of water resources. It can be asserted that the logic of opting for hierarchy water rights structure is because the costs of water allocation by negotiation are too high and so it turns to enforced administrative control to solve the problem of collective action in the allocation and utilization of water resources. That means the water rights hierarchy structure is also a product of economizing on cooperation costs and its effectiveness is preconditioned by the effectively lowering of management costs.

Water rights in the hierarchy governance structure is flexible and low quality of water rights is well matched with the administrative allocation and also with the lowering of management costs. As administrative means and market means are mutually exclusive, there is, theoretically, only administrative allocation in a pure hierarchy structure. Water rights hierarchy structure is an administrative allocation system, which uses administrative means to define and adjust rights, naturally incompatible with the market mechanism. Viewed from this angle, a hierarchy system that entirely uses market to allocate rights is a special case in the institutional hierarchy model constructed above, with its specialty lying in the adaptation of administrative means in all the allocation and reallocation mechanisms. Suppose allocation mechanism has been established at all levels, such specialty would be like that shown in Fig. 4.3. This has proved again that the institutional hierarchy model constructed by Challen (2000) is a descriptive method applicable in all multi-tiered structures for the management of natural resources and hierarchy structure is but a special case.

According to this understanding above, the water rights hierarchy concept advanced in this book is not very strict, because in the allocation mechanism, we have still retained the market means as an alternative option. It is out of the following considerations for this book to construct the water rights hierarchy model: first, it can completely exhibit the institutional option space for water resource management so as to make it easy to observe perceptually the multiple possibilities of water management system and enhance the special nature of China's water management system; second, theoretically China is a country with a hierarchical water rights structure and in practice market means may be introduced in some special circumstances. Market means is not only a theoretical possibility but also a reality in practice.

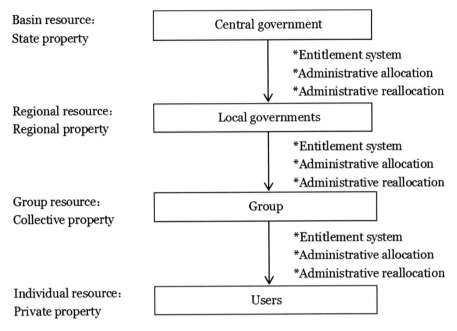

Fig. 4.3 Pure hierarchy water rights structure

4.4.3 Motivation Mechanism for Introducing Market Elements into Water Rights Hierarchy

As administrative and market allocation mechanisms are mutually exclusive, a pure hierarchy structure relies entirely on administrative means to allocate water rights and that is theoretically incompatible with market mechanism. However, when viewed dynamically, even a pure hierarchy system may also generate motivation for introducing market elements with the passing of the time. Under the theoretical framework above, market allocation may replace administrative allocation to become a more attractive option because of higher cost-effectiveness. We may also say that in the new environment, administrative means increases costs, thus reducing effectiveness and is ultimately replaced partially by market means.

The hierarchy system excludes market in that the transaction costs of the use of market are too high while the costs of the use of administrative means are relatively low. In the allocation of water resources, administrative means can better satisfy the following requirements than market means: first, water is essential for life, living and production. What is concerned about its allocation are security, impartiality and social acceptability; second, water resource involves the interests of all people and its allocation needs collective action and large scaled collective action; third, water resource is more serious than other natural resources in terms of uncertainty and market failure and that requires the leading role of the government; fourth, concomitant of water allocation are interests conflicts and these conflicts have to be mediated through negotiation or enforced coordination. These aspects are especially outstanding in China and that has further raised the costs of market means, thus making the advantages of administrative means even more significant.

However, the cost-effectiveness of administrative means is not absolute. The disadvantages of administrative means are the lack of incentive mechanisms for economizing and managing water, which is inseparably interconnected with the low quality of water rights. Generally, the higher the quality of water rights is which is held by a decision-making entity, the stronger the incentives is for meticulous allocation and use of water. The power system in the administrative allocation is invariably associated with low efficiency. Although market allocation of water resources has high costs, it has its own advantages. It can more accurately reflect the competitive demands and make economic incentives to act upon every decision-making entity, thus bringing about high efficiency. If the efficiency brought about by market allocation can offset all the costs, including the solution to market failure, social acceptability and costs of the operation of the market, market allocation would display its advantages.

The characteristics of administrative and market means show that low efficiency concomitant of administrative allocation should not be a serious problem when the resource scarcity is not serious enough. But it would increase opportunity costs with the rise in the value of water resources due to resource scarcity and the administrative allocation of resources is likely to give way to market allocation. So, with the rise in water scarcity, market allocation would become more attractive. The rise in water scarcity is the most important driving force behind the introduction of market mechanism, but not the only driving factor. Market has become an attractive option due to changes in external environment, such as the aggravation of scarcity of water resources and possibly the increase in the heterogeneity of users, changes in the external institutional arrangements and technical progress, which cause the cost-effectiveness of market means to rise in the new environment.

From a wider perspective, the motivation for the hierarchy structure to introduce market elements comes from two sources: one is the rise in management costs and the other is the lowering of cooperation costs. A pure hierarchy system may operate well when the management cost is kept at a low level. However, with the passing of the time and changes in the environment, the management costs may rise, so that it would make it difficult for the hierarchical structure to continue. The hierarchy would lose its effectiveness and the equilibrium governance structure would move toward a lower degree of hierarchy. Driven by minimization of general transaction costs for maintaining the governance structure, it is a rational option for introducing certain elements of market. Even if the management costs are not rising, the equilibrium governance structure would move toward a lower degree of hierarchy, too, and certain elements of market would be introduced if the cooperation costs are lowered. The introduction of market elements is, in the final analysis, the demand for minimizing general transaction costs of maintaining governance structure, irrespective of where the motivation comes from.

The bottom level of the hierarchy structure has the greatest motivation power than the upper levels for introducing market mechanism, because there are more decision-making entities at the lower levels and therefore have strong heterogeneity and the cost of administrative means is high. The higher the levels are, the smaller the number of decision-making entities are and therefore there would be greater homogeneity and lower costs of administrative means. When the external environment changes to cause the costs of administrative means to rise or the costs of market to lower, the lower levels of the hierarchy structure would be the first to introduce market elements. If the force of motivation increases, it would be transmitted to higher levels. This also implies that the higher the levels of the hierarchy, the more stable the hierarchy becomes and of the greater the opportunities to use administrative means. Conversely, the lower the levels are, the less stable the hierarchy is, with the bottom level being the least stable and that would make it easier to introduce market mechanism.

Once the market mechanism is introduced, the pure hierarchy structure would be knocked to pieces and the governance structure would move to seek a new equilibrium point at a much lower level. So the introduction of market elements is also an important manifestation of the reduction in the levels of the hierarchy of the governance structure. The higher the degree is, the bigger the scale is and the larger the scope is in the introduction of market mechanism, and the less the levels of the hierarchical structure becomes. This is, to a certain extent, also a manifestation of less levels in the hierarchy of the state governance structure.

4.4.4   Prerequisites for Water Market

Market allocation has always been excluded since ancient times in China's water management system. This is not unique in China. It is universal worldwide. Though water market is not rare in the world's history, for example, the water market in the Colorado State of the United States of America has a history of 150 years, large scaled introduction of market in water allocation happened only in the past quarter of a century and its spread as a new means of water management is only a thing of the recent decade. According to our logical framework, water market is rare in reality because its costs are too high. But how high, after all, is the cost of the use of market in the allocation of water rights? The World Bank listed the following nine prerequisites for successful operation of water market (Simpson & Ringskog, 1997):

1. *There must be a definable product to trade in the market. This product must be capable of being controlled, measured, and traded as a commercial good.*
2. *Demand for water must exceed supply. In many regions, sufficient water exists so that there are no competing demands.*
3. *The supplies derived from use rights must be transported to where the water is needed and be available when needed. Water flowing in a river during times of*

flood or times of monsoon rain has little value for agricultural, municipal, or industrial use and, instead, generally represents a detriment to the overall system. However, the same water stored for use when rivers dry up becomes a valuable commodity. By the same token, water stored during wet years to stabilize supplies during drought years has a much greater economic value than water flowing freely in the river.

4. *Buyers must feel confident that they will receive and be able to use the right purchased. The level of such confidence is reflected in the value of the right. For example, in capital stock markets, this confidence takes the form of an elaborate system of regulation, registration, and oversight. The stock exchanges are regarded as some of the most open markets in the world but still require a great deal of regulatory oversight. This is also the case with water use rights.*

5. *The water rights system must also resolve conflicts, because disputes between the use and ownership of water use rights always seem to develop. Historically, conflict resolution has taken the form of peer group resolution, administrative arbitration, and recourse to the judicial system. This conflict resolution process must be viewed by the market as fair and impartial and must be capable of effective and timely action.*

6. *The system must also apportion supply during periods of shortages and excess.*

7. *Well-defined and enforced mechanisms and criteria must be in place to assure that users are adequately compensated when their rights are confiscated or transferred to higher societal preferences. Markets for water use rights can function efficiently for voluntary redistribution, particularly when a relatively free and efficient system of markets exists for the rental of surplus supplies.*

8. *It is crucial in gauging the potential acceptability of water markets that the cultural and societal values of water resources be considered. Traditionally, throughout the world, many societies view water as a gift of nature or a gift of God, not subject to control, allocation, or dominion. In most countries, the legal ownership of a nation's water resources resides with the sovereign.*

9. *For any management program, including a market-based system, to succeed in the long term, it must be financially sustainable.*

This shows that market allocation of water rights does not only have prerequisites but also requires the establishment of a water market system, which is very costly. After summing up the experience and lessons of the countries with water markets, the World Bank arrived at this basic conclusion "although using the market process to reallocate water to meet changing demands is an important tool of water management, it is, in fact, just one tool in the process. It does not supplant education, public information, a strong hydro-meteorological database, strong administration and enforcement of water use rights, or a strong legal and institutional framework. When supported by all of the above, a market process will assist in achieving the highest and best use for water resources" (Simpson & Ringskog, 1997, pp. 4–7). This shows that the effectiveness of the water market depends on fairly low operational costs. This requires not only the lowering of the internal operational costs but also a well-matched external environment to further lower the operational costs.

The transition from administrative allocation to market allocation of water rights does not only requires static transaction costs for the operation of the new system but also dynamic transaction costs for institutional change. Examples show that the introduction of water market into an administrative system does not only require high operational costs but also high costs for the transition from the old system to the new. In developed market economies like the United States and Australia, which have a mature market environment and trading traditions, the costs of introducing market are low and it is comparatively easy to establish a water market system and may go even further in leaving the allocation of water resources to the market forces. But it is more difficult for developing economies to do so as their market environments and legal systems are not sound enough and the costs of the change are relatively high and the water market would display limited roles within a limited scope.

Based on the above discussions, this chapter concludes that there are two conditions that make the market work in reallocating water rights: one is that the dynamic transaction costs of introducing market mechanism must be low enough to overcome 'path dependence' of institutional change and ensure that institutional change take place; the other is that the static transaction costs must be low enough to make market operate normally (Bennett & Elman, 2006; E. G. Furubotn, 2005). But for China, which is used to employ administrative means to allocate water rights, it would be more difficult to introduce water market. The large scaled introduction of water market would entail very high dynamic and static transaction costs. If water market is made to display its role effectively, it would inevitably require the lowering of the two kinds of costs to a much larger extent.

Chapter 5
Water Rights Structure and Its Economic Explanation: Empirical Study of the Yellow River Basin in Ancient China

This chapter uses the hierarchy theory advanced in previous chapters to study the change in the water rights structure of ancient China. The Yellow River basin is the birthplace of China's agricultural civilization. Irrigation projects were started in the central Shaanxi Province before the Qin Dynasty and have lasted for more than 2000 years. Using this basin as a site for empirical studies, this chapter uses the water rights hierarchy conceptual model to describe the evolution of the water rights structure in ancient China in three periods of Qing-Han, Tang-Song and Ming-Qing. Then, it uses the theoretical knowledge provided in the previous chapters to explain the changes in ancient water rights structure in the following most outstanding four aspects: (1) Why did the state power gradually phase out of the micro management of irrigation affairs? (2) Why was the entitlement system in irrigation water allocation changed from input quota to resource quota? (3) Why did the initial allocation mechanism develop from license application to the system of registration? (4) Why were more and more market elements introduced into the re-allocation mechanism? This is a test of the hierarchy theory, which, during the course, is further enriched and developed.

5.1 An Overview of Water Resources in the Yellow River Basin

The Yellow River is the mother river of the Chinese nation and the birthplace of China's agricultural civilization. During the more than 3300 years of history from the Xia Dynasty in the twenty-first century B. C. to the Northern Song Dynasty, the Yellow River basin was a political center as well as an economic center of China, which did not move south of the Yangtze River until after the Southern Song Dynasty due to wars and ecological changes. Even so, the Yellow River basin maintains a significant position in the political and economic activities.

© Springer Nature Singapore Pte Ltd. 2018
Y. Wang, *Assessing Water Rights in China*, Water Resources Development and Management, DOI 10.1007/978-981-10-5083-1_5

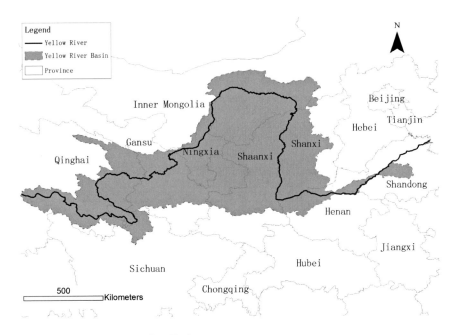

Fig. 5.1 Map of the Yellow River Basin

Historically, droughts and floods were frequent in the river basin, especially in the lower reaches. The 2540 years from 602 B. C. to 1938 witnessed 1590 floods and more than 20 changes of the main river course, bringing about untold miseries to the Chinese nation. The river changed its main course many times, especially in the lower reaches. The current main course was the result of the 1855 floods at Tongwaxiang. Now the river flows 5464 km, through Qinghai, Sichuan, Gansu, Ningxia, Inner Mongolia, Shaanxi, Shanxi, Henan and Shandong, covering a catchment area of 795,000 km^2 (See Fig. 5.1).

5.1.1   Characteristics of the Water Resource of the Yellow River

The natural run-off of the Yellow River basin is 58 billion m^3. This, plus 13.9 billion m^3 of ground water (not including the volume repeated in calculation), brings the total volume of water up to 71.9 billion m^3. The volume of water varies greatly between years and the distribution of water volume within a year is concentrated and uneven from area to area, which is common in all rivers in North China. At the same time, it carries plenty of sands, coming from different sources, which is quite different from other rivers.

First, the flow is small and filled with sand. The natural run-off averages only 2% of the total run-off of all rivers in the country for years. The per capita water volume in 1997 was only 543 m^3, about 25% of the national average. The per-*mu* (1/15 of a hectare) water volume averages 307 m^3, about 16% of the national average. The average amount of sand carried downstream is 1.6 billion tons a year. The sand contained in the river run-off averages 35 kg/m^3. The trunk river course contains 920 kg/m^3 of sand on average. It is a river with the biggest amount of sand content in the world. The sand deposits on the lower reaches to raise the river bed, so much so that the river is known as a suspension river. Small volume of water but big volume of sand makes it difficult to tame.

Second, the volume of water changes greatly between years and distribution is concentrated and the low water level period lasts long within a year. The biggest run-off at all hydrological stations is usually 3.1–3.5 times the minimum annual run-off in the main course and 5–12 times in the tributaries. The total run-off of the main course and major tributaries during July–October flood season is more than 60% of the annual total. The volume of run-off during non-flood seasons is limited, often causing water shortages and dry-up in sections of the river. There have been three consecutive periods of low water level since the field survey data available, with the one in 1922–1932 lasting for 11 years and that in 1990–2002 lasting for 13 years.

Third, water and sand come from different sources and distribution of water is not in conformity with that of land resources. Most of the run-off comes from sections above Lanzhou, accounting for 56% of the total and more than 90% of the sand come from the middle reaches, with about 900 million tons of sand washed downstream in the Hekouzhen-Longmen section, 56% of the total sand carried by the whole river. The Yellow River basin and the irrigated areas on the lower reaches have abundant land resources, but most of the cultivated land is concentrated in the arid riparian areas of Ningxia, Inner Mongolia and the Fenhe River and the Weihe River valleys and the run-off short irrigated areas on the plains of the lower reaches.

5.1.2   History of the Development and Utilization of Water Resources of the Yellow River

The Yellow River basin was one of the earliest areas to develop irrigated agriculture. Artificial irrigation probably started in the later period of the Spring and Autumn period, first by digging wells and then by diverting water from rivers. The most famous water diversion project recorded in history was built by Xi Men Bao during the Warring States period, who ordered 12 canals dug to divert water from the Zhanghe River to the Town of Ye to irrigate the farmland (Twitchett & Fairbank, 1978). From then on, all the dynasties including Qin-Han, Tang-Song, Yuan-Ming and Qing built a large number of water conservancy projects in the Yellow River basin. Agriculture, especially in Ningxia, Inner Mongolia, the middle

Table 5.1 Development and utilization of the water resources of the Yellow River basin (1949–2014)

	1949	1980	1990	1999	2003	2014
Total population (million)	37.00	81.53	97.81	112.52	na	na
Irrigated area (million ha.)	0.80	4.36	6.00	7.53	na	na
Total volume of water drawn (billion m^3)	9.0	35.8	47.8	51.7	42.9	53.5
% of water used in agriculture in total (%)	97	88	85	82	74	67
% of surface water drawn in total drawn (%)	96	77	76	74	69	77
% of water drawn from surface water for use in agriculture (%)	99	94	93	91	81	73

Sources: Data for 1949 are estimates by the author; data for 1980 are calculated based on the research achievements in the study of the Yellow River resources by using economic model; Population data and data for 1990 come from the achievements of the project of studying "Yellow River Water Control and Development and Utilization" in the Eighth Five-Year Plan Period; data for 1999, 2003 and 2014 come from "Yellow River Water Resources Bulletin". Irrigated area include those irrigated by both well water and water drawn from the lower reaches of the Yellow River. Population in the Yellow River basin includes the population of the four cities of Zhengzhou, Kaifeng, Jinan and Dongying on the lower reaches of the Yellow River

of Shaanxi and the Fenhe River basin in Shanxi, was the main consumer of the Yellow River water. But irrigation development was limited by technical and economic conditions. A total of 800,000 hectares of farmland were brought under irrigation after more than 2000 years of development using only about 9 million m^3 of water from the river (see Table 5.1).

Since its founding, the P.R. China has unfolded a large scaled water development and utilization drive in the riparian areas of the Yellow River. In a short span of 8 years from 1949 to 1957, the country brought about 670,000 hectares of more farmland under irrigation, about the total for the previous thousands of years. After the 1970s, more work was done. Now, there are more than 3100 large, medium-sized and small reservoirs in the Yellow River basin, with a total holding capacity of 57.4 billion m^3. In addition, there are now more than 4600 water diversion projects and 29,000 water lifting projects. On the lower reaches, 245 water lifting projects have been built to supply water to the Yellow-Huaihe-Haihe plain area. Now the Yellow River, which has only 2% of the river run-off in the country, is irrigating 15% of the farmland and supplying water to 12% of the national population and more than 50 large and medium-sized cities. The total areas of farmland irrigated by the Yellow River water have reached more than 7.5 million hectares, nearly ten times as much as at the beginning of the founding of New China (see Table 5.1).

According to the 2014 Yellow River Water Resources Bulletin, 53.298 billion m^3 of water, including surface water diverted from the river basin to other areas, were withdrawn from the Yellow River in 2014, which also includes 40.476 billion m^3 of surface water, accounting for 77% of the total. The total water consumption was 42.675 billion m^3, including 33.187 billion m^3 of run-off, accounting for 79% of the total. The total amount of water consumed by agriculture in the year was 28.941 billion m^3, accounting for 67% of the total, with surface water accounting for 73%.

Table 5.2 Current Yellow River water consumption (2013) (Unit: Billion m^3)

Water	Agriculture	Industry	Urban and rural daily life	Total
Surface water	24.255	3.424	1.503	29.182
Ground water	4.686	1.849	1.239	7.774
Total	28.941	5.273	2.742	36.956

Source: "2013 Yellow River Water Resources Bulletin"

Water for industrial use accounted for 12%, of which surface water made up 65%. Water consumption by urban and rural population was 6%, of which surface water made up 55% (see Table 5.2).

5.1.3 Historical Evolution of Water Resources Management in the Yellow River Basin

Floods were frequent in the Yellow River in ancient times. The main task of river control at the time was to prevent floods. Since the very beginning, the river course on the lower reaches was put under the direct control by the central government. The Eastern Han Dynasty commissioned water control officials. The Tang Dynasty set up river control headquarters under the Ministry of Projects. The Song Dynasty expanded river control organizations. The Jin Dynasty commissioned officials to patrol river and all the prefecture and county magistrates were charged with the tasks of preventing floods. The Yuan Dynasty commissioned water superintendents and the head of the Ministry of Projects was the chief of water control. The Ming and Qing dynasties appointed officials to coordinate inter-regional and basin-specific water control projects. In the modern nationalist government period, there was a Yellow River Council to operate water projects on the Yellow, Weihe and Luohe rivers. In 1946, the Heihe-Shandong-Henan liberated area set up a Yellow River Conservancy Committee. At the beginning of 1950, the State Council set up the Yellow River Conservancy Commission, which has been expanded and improved in function over the past half century (Chen, 2002). The history of the Yellow River water resource management may be divided into three periods.

The first period spans from the Qin Dynasty to before the founding of the P.R. China. In the prolonged ancient society, due to underdevelopment of the economy, the water resources available in the Yellow River were far more than what social-economic development needed. During this period, the demand for water could be easily satisfied by lifting natural water directly from the Yellow River or by building some simple projects. Water resources management was limited to some given water projects. There were no inter-regional negative externalities. The main factor restricting water use was scarcity of projects due to limited technical and economic conditions. The allocation of water rights was done mainly at the group-user level. But resource scarcity in local areas caused disputes in water rights allocation among different groups.

The second period covers the years from the founding of New China to the planned economy period before the introduction of reform and opening up policy. After New China was founded, the demand for water in the Yellow River basin increased rapidly and so did water diversion projects. This led to a serious tendency of stressing construction to the neglect of management during this period. Water management was so weak that it was confined to some given water supply projects. Allocation of water rights was still done at the group-user level. The unbridled withdrawing of water from the river caused resource scarcity in some local areas during dry years and dry seasons. Water resource allocation mechanism had to be introduced in a larger scope. Artificial intervention had to be introduced from the central government level to the local government level, and from the local government level to group level in the development and utilization of water resources. Surveys and evaluation of water resources as well as water resources planning also started.

The third period is the transitional period from the planned economy to market economy since the introduction of reform and opening-up policy. Beginning from the 1980s, resource scarcity appeared in the whole of the Yellow River basin, with water available falling far short of demand in local areas. Disputes over water use began to escalate. Water resources development had to be shifted local governments to the hands of basin organization. This is a period of large scaled institutional change. In 1987, the State Council mapped out a scheme for the allocation of water available in the main course of the Yellow River; in 1988, it went on to promulgate the Water Law, followed by the introduction of the water withdrawal license system and a series of policies and regulations concerning water resource management. Basin organizations began to exercise water allocation power on behalf of the state. A basin-wide management system supplemented by regional management took shape. Water rights allocation mechanisms were established at the group and local government level, which, though undergoing rapid changes, is still in the process of improvement, acquiring many new characteristics.

As is shown in Table 5.1, surface water has remained the main part, about three quarters, of the total water withdrawn from the Yellow River, although it is assumed a downward trend. The ground water abstraction is on the rise, as ground water abstraction mainly has some regional impact, its allocation mechanism is different from surface water. The surface water flows basin-wide and its use has an impact in the whole basin. It is the main target of management and also the object of basin-wide water allocation. As agriculture is the main user of surface water, the following sections will mainly study the utilization of surface water in the Yellow River basin, without touching on the use of ground water.

With the progress in industrialization and urbanization since reform and opening up, water use in the Yellow River has kept rising and the water use structure has undergone constant changes. The general trend is characterized by continued drop in the water used by agriculture but steady rise in the water used by non-agricultural activities. The total amount of water used by agriculture dropped to 67% of the total by 2014. Although water used in irrigation assumed a downward trend, agriculture remained the main user of water, mostly surface water. In 2014, the surface water used by agriculture made up 73% of the total in the whole river basin (see Table 5.2). This chapter focuses on the use of surface water by agriculture.

What needs explaining is that, due to the vast territory covered by the Yellow River and variations of geographical, climatic and hydrological characters in different regions and in the habit of using water, the irrigation area management and water use management are not entirely the same. But fortunately, the social structure since ancient times has determined the strong homogeneity in the public affairs management model among administrative regions at all levels, especially in the Yellow River basin. So the study of typical cases also reflects the picture of the whole basin.

5.2 Changes of Water Rights Structure in Ancient China

5.2.1 General Description

The Yellow River water was mainly used to irrigate farmland in ancient times. According to this author's estimate, 96% of the water withdrawn from the Yellow River was surface water and 99% were used by agriculture (see Table 5.1). Due to limitations by the then productivity level, irrigation was limited in scale. Irrigated areas were mainly concentrated in the upper and middle reaches of the river, including the River Bend area and such tributaries as Fenhe, Weihe, Jinghe, Luohe and Huangshui, where there were no dams built and irrigation depended on natural flow.

In the ancient society, the supply of natural water was much bigger than demand. Project scarcity remained the main factor accounting for water resource scarcity and water rights issues were confined to the scope of water supply projects. Water rights allocation was mainly the allocation of water rights for irrigation. Water rights were mainly allocated at the group-user level. The water resources in the basin as a whole and at the regional level were in a state of open access. Water allocation among localities and among groups only appeared in local areas. The disparities in the water rights structures during different historical periods were mainly manifested in the changes of the mechanisms for allocating water for irrigation.

The general principle for allocating water for irrigation in the whole ancient society was equality. The idea of equal distribution and equal benefit ran through all kinds of water laws and regulations and in local historical records. The principle of universal equality without any discrimination is laid down in what is known as *Shui Bu Shi*, China's first water law promulgated in the Tang Dynasty. The rules for providing the sequential order of the rights for irrigation were for the purpose of ensuring that the whole irrigation area was equally benefited. The "*Da Tang Liu Dian*" (the six codes of the Tang Dynasty) states that rice should be irrigated first and arid crops should come later, all the water use must come from the lower reaches, and even at times when water was not enough, the method of 'irrigation by turns' was adopted (Twitchett & Fairbank, 1978).

The egalitarian idea was originated in the Confucian culture. 'Inequality rather than want is the cause of trouble'—that is the tradition. Obviously equal allocation was readily acceptable and most favorable for maintaining the order of irrigation. Guided by such an idea, irrigation water was mainly allocated according to the quantity and grades of land. Although the irrigation systems of all dynasties were different, 'to fix water quotas according to the size of land' remained the basic water allocation principle (Gu, 1997).

The change of dynasties had a profound impact on farmland irrigation. When a dynasty was on the decline, water projects would decline, too. In the period of peace, there would be a big drive for water conservancy projects. Only when there was pace and stability, was it possible to have a sound irrigation system. Once the country was plunged into turmoil, the system set up would be sabotaged or interrupted until a new dynasty came into being. Though the more than 2000 years from Qin-Han to Ming-Qing witnessed numerous twists and turns in the allocation of irrigation water, it experienced the stages of startup, development and flourishing. This study has selected three periods of the ancient times as typical in the evolution of irrigation water rights allocation. The Qin-Han period was the initial period when water allocation just began; the Tang-Song period marked a high degree of development; and the Ming-Qing period marked maturity of the allocation system.

5.2.2 Qin-Han Period

Qin-Han was a period when the ancient feudal society was established and developed and also a period for the initial establishment of irrigation management system. Farmland irrigation and management in the period showed the following characteristics:

First, it showed high dependence on the government in building water projects. During this period, the state committed a large amount of financial resources to digging ditches and reclaim land. The central government was directly involved in building major irrigation projects. Wu Di of the Han Dynasty ordered Ni Kuan to take charge of the work of building six auxiliary canals (Liu Fu Qu). Irrigation projects and related affairs were left to local governments. Local officials were often charged with direct responsibility for building irrigation works. Li Bing, an official in Sichuan of the State of Qin led people in building the sophisticated irrigation works known as Dujiangyan. Zhao Xinchen, an official of Nanyang in the Western Han dynasty, oversaw the building of a large number of irrigation projects (Twitchett & Fairbank, 1978).

Second, appointing full-time official to take charge of major irrigation works. According to historical records, the eastern Han Dynasty commissioned a superintendent to take charge of irrigation works and all the irrigation projects had their

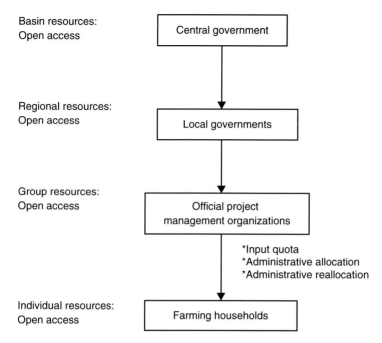

Fig. 5.2 Surface water rights structure of Qin-Han period

own management rules. The earliest irrigation rules were recorded at the beginning of the Western Han Dynasty when the six auxiliary canals were dug. When Zhao Xinchen was building irrigation works in Nanyang in the Western Han period, he formulated rules on the restricted use of water and engraved them on a stone tablet, admonishing people to economize on the use of water and use the limited resources to irrigate more farmland (Gu, 1997).

Third, it was governed by crude irrigation management rules. As the means for allocating and measuring water was still underdeveloped, water quantity was by the rule of thumb at the beginning. Such an ability was unable to allocate water quantity directly. So prohibitory rules had to be employed to achieve the purpose of economizing on the use of water (Chi, 1936).

In the Qin-Han period when, it was not long after artificial irrigation started, the state began to make irrigation management rules while building irrigation works. The water rights structure during this period is shown in Fig. 5.2. Irrigation water was then allocated under the sponsorship of project management officials. All the water rights allocation and reallocation were done by administrative means. Mainly the input quota entitlement system was used. The central and local governments were not directly involved in the allocation of water rights. What they did was to organize the building of irrigation projects and commission officials to manage irrigation works.

5.2.3 Tang-Song Period

Tang-Song period marked the heydays of the development of China's feudal society and also a period of big development in irrigation. During this period, there were some new features in the building and management of irrigation projects.

First, the irrigation management system was brought to near perfection, with management organizations set up at all levels from central down to grassroots. The central government appointed full-time officials to manage farmland irrigation. Sui, Tang and Song dynasties all had water departments under the project ministry to manage and oversee irrigation projects. What was known as *"Du Shui Jian"* established during the period was a full-time technical organizations charged with planning, engineering and management of irrigation projects. The irrigation area was managed by officials sent down by the central government. There were specific rules on the organizational setup of irrigation area management organizations. Even the grassroots management personnel were appointed by officials. Officials were in absolutely dominant positions in the management of irrigation areas. In the Tang Dynasty officials were directly involved in the installation of water measurement facilities and the formulation of rules for the allocation of irrigation water.

Second, irrigation management gained unprecedented development. The Tang Dynasty promulgated the Water Law, the first of its kind in Chinese history. During the reform period of the Northern Song Dynasty, Wang Anshi promulgated *"Restrictions on Farmland Irrigation"*, which was the first fairly complete farmland irrigation law in China. The *Water Law of the Tang Dynasty* covered a wide range of contents, with detailed regulations on the use of irrigation water. This reflected the high level in farmland irrigation management in the period.

Third, the state investment in irrigation began to be reduced after the Tang Dynasty. The money needed for building irrigation works had to be shared by water users, who were also obliged to pay irrigation area operational fees or pay in labor or in grain to get the rights to use water. This became a common practice in later dynasties. Besides, in the literatures of the Song Dynasty, there were also records on the government encouraging people to build irrigation projects.

Fourth, great development was made in water allocation technology, making it able to use sluice gates to regulate the proportion of water from the main river courses and tributaries. The system of water withdrawing license application was then introduced. Canal or ditch management personnel applied for the permission to withdraw water according to the crop areas and quantity of water needed. The *Shui Bu Shi* states that the areas of crops must be known before applying for water withdrawing license. The application for water withdrawing license is very much like the water ticket system in the modern times, when progress in water allocation technology and institutions made it possible to allocate water according to quotas allocated among different canals within a waterway system (Twitchett & Fairbank, 1978).

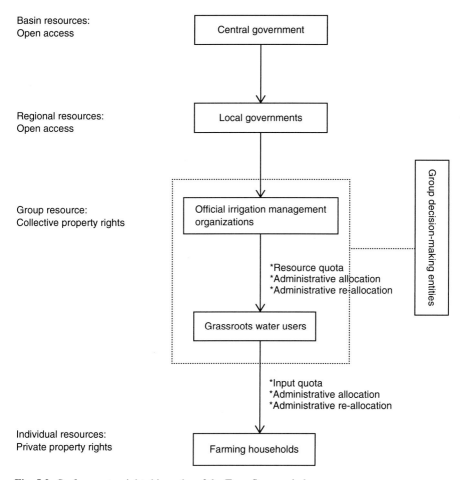

Fig. 5.3 Surface water rights hierarchy of the Tang-Song period

The official institutions governed by law were brought to perfection in the Tang-Song periods. The state control over irrigation reached its peak in the ancient society. The water rights hierarchy in this period is shown in Fig. 5.3. Compared with the Qin-Han period, the central government's functions began to develop toward institutional construction and macro management. Local governments set great store by managing irrigation projects. Within irrigation areas, there already were complete mechanisms for allocating water rights administratively. The entitlement system began to appear in the form of resource quota. With the increase in small irrigation works built by non-governmental forces, non-governmental forces began to participate in water management.

5.2.4   Ming-Qing Period

Ming-Qing was a period of decline in the ancient feudal society and also a period of maturity in the irrigation management. During this period, new developments were made in farmland irrigation on the basis of what was achieved in previous dynasties.

First, the number of non-governmental independent forces increased. After the Yuan Dynasty, management personnel of grassroots irrigation areas were no longer directly appointed by the government, but were democratically elected or recommended. Till the Ming and Qing dynasties, grassroots irrigation management system was combined with neighborhood communities, bringing the autonomous management level to a historical high. The state promulgated few laws on irrigation but carried forward what was established in the previous dynasties and made them established common practice, which was enforced by virtue of the neighborhood communities and traditional ethics. This was known as non-official system represented by the neighborhood rules and agreements, which played a tremendous role in irrigation management. Besides, what merits attention is the increase in small irrigation works run by non-governmental groups, which are jointly run by the government, non-governmental groups and people under the supervision by the government. The right of initial irritation water use was guided by the principle of equality, which was expressed in a given volume of water for irrigation. The water volume was measured by the water depth per unit land, which took "*li*" as the unit. It was difficult to operate in real irrigation. It had to be converted by associated methods. In part of the time during the Song, Yuan and Ming dynasties, the water volume was measured by what was known as 'Jiao', a cross-section of water passage. One *jiao* is one *Chi* in breadth and one *Chi* in depth, and one *Jiao* of water may irrigate 80 *mu* of land in 24 h. Management personnel used *Jiao* as a unit to measure the water volume in water allocation. This method evolved into time measurement up to the Ming and Qing dynasties, that is, water rights were allocated according to the duration water passed through a ditch, which was called 'water path' (*Shui Cheng*). The water path is measured by burning incense, a method widely applied in the Ming-Qing period (CIC, 2005).

Third, water withdrawal license application system evolved into a water registration system. The registration book was established under the oversight by officials. The water was allocated according to the area of land. So the basis of water allocation remained the right to the land. Once the registration was fixed, it in fact became land registration and so had the nature of local laws, which was kept stable for a long time. It is the beginnings of the modern water rights registration system. This system features long stability, transferability and inheritability of water rights, assuming the possibility of assignment.

Fourth, water rights trading already took place in part of the irrigated areas, often covertly, despite the ban imposed by the state from Tang through the Ming and Qing dynasties. It is underlined that land is land and water is water, and land and water are separated in trading. Water is priced arbitrarily, while land can be sold

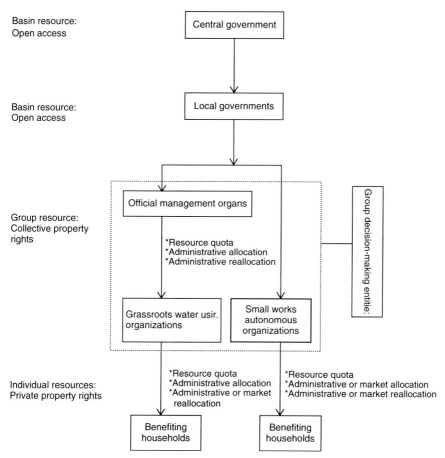

Basin resource:
Open access

Basin resource:
Open access

Group resource:
Collective property
rights

Individual resources:
Private property rights

Fig. 5.4 Surface water rights structure of the Ming-Qing dynasties

separately. Such trading was true in the Guanzhong irrigated area (Liu, Duan, & Deng, 2014a). De facto water rights trading appeared in the Ming and Qing dynasties on the then given historical conditions and as the result of changes in technology and institutions. While the state maintained its controlling power of irrigation, non-governmental forces also developed, which, together with state control, brought the irrigation management level to the peak. The water right structure of this period is shown in Fig. 5.4. Compared with the previous dynasties, the importance of the central government in irrigation project and institutions supply dropped greatly. But local governments still played an important role. The main river course and tributaries were still subject to official management and those below the sluice gates were subject to management by the people. There were more small irrigation works built by local people and autonomous management was

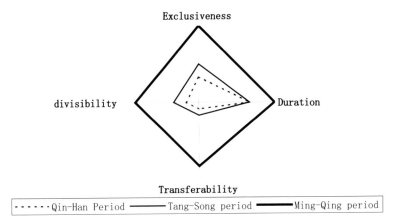

Fig. 5.5 Comparison of the characteristics of irrigation water rights during different periods of ancient China

expanded in scope.[1] Resource quota entitlement system was adopted in the period. However, water rights of small irrigation works were extensively allocated by market.[2] Water rights may be transferred through the market in re-allocation.

5.2.5   Characteristics of Irrigation Water Rights in Ancient China

In general, there were great uncertainties in the allocation and adjustment of irrigation water rights in ancient China. It was a highly incomplete private property right, resulting from the then social structure and the historical background of the rise and fall of dynasties. Though the evolution was slow over the past 2000 years, big progress was made in the management system of irrigation water rights, with the allocation mechanism becoming more and more perfect and the characteristics of water rights changing gradually. On the basis of the study of the water rights structure in the three typical historical periods, Fig. 5.5 sums up the characteristics of the irrigation water rights during different periods from the perspectives of exclusiveness, duration, transferability and divisibility. As is shown in Fig. 5.5, the characteristics of water rights from the Qin-Han to Ming-Qing were all

---

[1]The management personnel of small irrigation works included Canal Chief, Conferdam Chief, Headmen, Association Head and the Eldermen, all recommended by water users or appointed by turn.

[2]In small irrigation works, farmers got their initial irrigation water rights according to their shares of investment in the projects.

enhanced in the four characteristics, indicating the rising trend of the quality of water rights.

The evolution of water rights characteristics is the result of technical progress and institutional changes. In the Qin-Han period, the technology in allocating and measuring water was underdeveloped andunable to allocate water among different river systems. The irrigation water was shared in the whole of an irrigation area. Water use was highly dependent on other users, with a very weak exclusiveness. Up to the Tang-Song period, thanks to the development of the techniques for allocating water, water could be measured within an area covered by a sluice and a water withdrawal license application system was introduced and water was allocated and used in a planned way during irrigation seasons, thus enhancing the exclusiveness of water rights. Up to the Ming-Qing period, water path technique was widely applied and the technical progress gave rise to the water registration system, with water rights officially registered. This was widely accepted and observed. Due to inheritability and transferability and even trading, the exclusiveness of water rights, the fundamental feature of property rights, was enhanced greatly. With the rise in the degree of exclusiveness, the other characteristics were also enhanced correspondingly. Figure 5.5 shows exactly this point.

5.3 Logic of Evolution of Ancient Water Rights Structure

5.3.1 Evolution of State Roles in Ancient Water Rights Management

State in ancient China is referred to the political system of empire. In the hierarchy conceptual model in this book, the central and local governments were the main bodies of the state polity while groups acted as a bridge between the government and people, or a transitional form from the government to the people. The state force in the hierarchy governance structure was always extremely powerful and in a dominant position while the rights of the people were always limited, subject to the administrative powers of the state. Study of the roles of the state in water rights management is a focal point in observing the changes of water rights structure and in making up for what is lacking in the description by the water rights hierarchy conceptual model.

The state regulates the utilization of irrigation water in the following four aspects. First, to decide on the basic principles for the allocation of water resources, Xiao pointed out that the principle of water allocation in the Guanzhong area was fixed by the state at the very beginning and it had never been replaced although individual petty farmers, local country gentlemen and power families all had their influence in the course of allocating water. Second, to oversee technically the irrigation system and water use methods, the state, from the very beginning, was involved in the oversight of farmland irrigation in the Guanzhong area and often

codified the technical measures. Third, to participate in the maintenance of irrigation projects and the organization and management of irrigation processes, the state
did not only exercise technical oversight, but also played an important role in
maintaining irrigation projects, organizing and managing the use of water. All the
dynasties following Tang commissioned officials in the Weibei irrigation area,
together with local officials, who managed projects and handled other affairs
concerning irrigation. Fourth, to mediate and pass ruling on water rights disputes
according to law, the arbitration procedures concerning water rights in ancient
times show that the realization of water resource use right could not go beyond
the scope provided by law and regulations. As the owner of water resources, the
state power always played a dominating role in the use of water (Gu, 1997;
Twitchett & Fairbank, 1978).

But the ways and characteristics of the state varied in different historical periods.
According to the study of the changes of ancient water rights in China by this
author, the state's role in water rights management may be divided into three typical
stages, which are corresponding to the three historical periods mentioned above.
The state's role was seen mainly in irrigation project construction during the Qin-
Han period; its role in the Tang-Song period was mainly in the institutional
arrangements; and its role in the Ming-Qing period was shifted to macro regulation
and control. The general description of the changes in ancient water rights structure
mentioned above already shows such trend. The following is a description of the
nature of such changes.

Large scaled irrigation drive started in the Warring States period in ancient
China. Under the historical conditions, development of irrigation was vital in
making the country strong and powerful and the state played a dominant role in
building irrigation projects. The most famous irrigation projects built by the Qin
Dynasty included Dujiangyan in Sichuan, Zhengguo Canal in Central Shaanxi and
Lingqu in Guangxi. The Han Dynasty, which lasted a fairly long period of time,
built irrigation projects in all the Yellow River basin, such as Huangshui in the
upper reaches, Luohe, Weihe and Fenhe basins in the middle reaches and Wenshui
water diversion project in the lower reaches. The irrigation projects built under the
sponsorship of the state in the Qin-Han period mainly served military purposes and
the needs of the rulers. As technical and economic conditions were poor, the main
tasks were to expand the supply of irrigation projects to overcome project scarcity
and irrigation management was naturally in a subordinate position. That is why the
irrigation management in the Qin-Han period was only regional or local. For
instance, Ni Kuan of the Western Han Dynasty built six auxiliary canals for
irrigation. Zhao Xinchen organized the building of irrigation projects in Nanyang
and imposed restrictions on the use of water by the people. Historical materials
available show that there were both lack of irrigation management system and
records on micro management activities and water rights system (Chang, 2001). It
can be understood that as the state was devoted to project building, irrigation
management and water rights management were rather crude.

With the economic and social development in the heydays of the Tang Empire,
irrigation also gained some ground, especially in the institutional arrangements

with regard to irrigation system. The central government issued the most famous *Shui Bu Shi*, which was China's first water law promulgated by the central government. The law provided for all aspects of irrigation affairs, with very detailed provisions on irrigation management. While stressing the building of irrigation projects, the Song Empire also began to heed water resource management. The imperial court issued a series of orders for developing water resources and during the Wang Anshi reform period, issued 'Rules on Farmland Irrigation', which was a fairly complete policy concerning farmland irrigation and resulted in a most famous irrigation upsurge in the Chinese history (Gu, 1997; YRHEO, 1995). Tang-Song period achieved glorious achievements in irrigation management, which may be regarded as the zenith in the Chinese irrigation history. The state played the role of institution supplier. However, in practice, the state also got involved in specific management of water affairs, indicating the tendency for the government to have an oar in all affairs.

The state role in irrigation began to decline in the Ming-Qing period, in which the state had little to do in institution supply, giving way to non-official local folk rules and regulations, which partially replaced the role of the state. According to the study by Xiao Zhenhong, (starting from Tang) state power organizations exercised management over water rights very specifically and to the minutest details, such as the Tang Dynasty imperial water superintendent and prefecture and county officials all had their hands in irrigation affairs. In the Song-Yuan period, grassroots management personnel must specify the area of land and crop types in the water withdrawal license application materials. After Yuan, local governments tended to give up management in the micro level. Up to the Qing Dynasty, the county government no longer handled water rights allocation with a river basin except water rights law suits (CIC, 2005). This author reaches the conclusions above and holds that after Tang and Song, there was indeed the tendency of giving up micro management of irrigation affairs and the state roles were limited to the support by common law, definition and adjustment of water rights, water rights law suits and other aspects of macro control and management. The narrowing of the scope with state intervention was closely associated with the growing non-governmental forces dominated by groups.

5.3.2   Changes of the Role of the State in Water Rights Management

It is easy to understand the changes in irrigation affairs from the Qin-Han period to the period of Tang-Song. The low level of productivity led to constant internal conflicts and the marginal benefits of irrigation projects at the very beginning were very high. When project scarcity prevailed and there was a lack of voluntary association, the state's input and marginal benefits were relatively low. In such circumstances, it became a rational option for stressing the building of irrigation projects to the neglect of management. When up to the Tang-Song period, large numbers of irrigation projects were built and all the crop fields were dotted with large and small irrigation works and the marginal benefits tended to grow by strengthening irrigation management and expand the areas under irrigation and the unified supply of irrigation management institutions instead of separated management institutions supplied by various irrigation areas may economize on management costs. So the state played an increasingly bigger role in institutional arrangements and that should also be regarded as a rational option.

What is more of interest is the question why the state role began to shrink from Tang-Song to Ming-Qing period? or more accurately, why did the state force gradually phase out of specific irrigation affairs and turned to macro regulation? The explanation by the hierarchy theory discovers that it was driven by lower management cost within the hierarchical structure. The following is an exposition of this point.

The study framework of this book indicates that the governance structure is the result of the minimization of transaction costs under environment constraints. In essence, the basic task of irrigation governance in ancient China was to resolve the conflicts between population and resources (especially cultivated land resources), raising land productivity and ensuring grain output compatible with the size of the population or ensuring grain security (Hu & Wang, 2000). This task required developing irrigation systems and raising the efficiency of the irrigation works.

The *Shui Bu Shi* of the Tang Dynasty formulated a series of rules for managing irrigation works. It was, in fact, the formulation of a complete set of irrigation management rules to effectively maintain the irrigation order and economize on management cost. *Shui Bu Shi* laid down the basic principle aimed at rational allocation of irrigation water (Gu, 1997), including egalitarian principle, the water withdrawal license application system, the water volume control at sluice gates and maintaining the water level of rivers so as to achieve the maximum natural flow of water for irrigation, irrigation area statistics, sequential order of irrigation areas and irrigation by turns. These institutions reflected the management level of the time. Compared with the management systems formulated by irrigation areas, the practice for the central government to issue unified orders to spread advanced institutions in the whole country at the minimum costs was in fact the low transaction costs brought about by the hierarchical structure.

The set of management institutions established by the Tang Empire were executed by state agents commissioned by the central government all the way down to the grassroots. From the installation of water allocation facilities to the formulation of allocation mechanism within river basin or irrigation areas were all undertaken directly by the government. People must apply for license to use water and the areas of all the irrigated land must be registered. Such application was done every year. The irrigation management system was very effective during the Tang Empire, thus resulting in an agriculture that was so developed that it had never been seen in previous dynasties.

The governmental and non-governmental forces in irrigation management were not mutually exclusive but mutually reinforcing each other. If an official agent commissioned by the state had direct relations with farmers, the costs would be very high. If the state had direct relations with farmers through its unofficial agents, the governance costs would be greatly economized. Further study has proved that even in the Tang Dynasty, it was not that state took everything to itself. The maintenance and management of the terminal canals were done by grassroots organizations formed by beneficiary households. But the non-governmental organizations needed the support of the government. In the irrigation area of Shazhou in the Tang Dynasty, there was an organization called 'Qu She' (canal society), formed by about 20 beneficiary farming households. The head of the society would lead the members to repair irrigation works with their own tools and materials when need arose. If any member was absent, that member would be punished according customary practice (CIC, 2005; Gu, 1997). In this example, the head of the society played the role as an unofficial agent of the state.

If we compared unofficial state agents to the link in water rights structure between the government and non-governmental organizations, the link in the Tang Dynasty was very close to farming households. However, such link began to move upward in the water rights structure until a large number of local irrigation organizations appeared in the Ming-Qing period, which were managed by local groups and whose personnel were selected by localities. The state did not have a hand in routine operations of the irrigation works. The state role was expressed in granting the local management rules a legal status and ensured that they were observed. When disputes occurred, the state would act as an arbitrator (Zhang, 2012). The changes in the power allocation pattern in the hierarchy resulted in the economizing of input in management, thus objectively lowering the costs of the operation of the hierarchical structure.

The phasing out of the state power from the grassroots irrigation management from Tang right down to the Qing dynasty was regarded as a natural evolution driven by transaction cost economizing and an inevitable option manipulated by the logic that hierarchical structure may lower management cost. From Tang to Qing, the shrinking of the state role was pre-conditioned by institutional changes, which manifests itself in the following three aspects: the way of entitlements developed from input quota to partial introduction of resource quota; the way of initial allocation developed from water withdrawal license application to water registration; and the way of water rights re-allocation developed from administrative means to market forces. The causes for such changes will be dealt with in the following sections. It is just these important institutional changes that made it possible for the state roles to shrink and for the management costs of the governance structure to become lower and for the performance of the whole water rights structure to improve.

5.4   Institutional Change of Ancient Water Rights Allocation

As is shown above, there were three trends of changes in the water rights structure in the prolonged ancient China: the way of entitlements developed from input quota to partial introduction of resource quota; the way of initial allocation developed from the water withdrawal license application to water registration; and the way of water rights re-allocation developed from administrative means to market forces. These are exactly the three major factors of allocation mechanisms of the water rights hierarchy conceptual model. Chapter 4 dwells on the economic logic for the option of water rights allocation mechanism. The following is an empirical analysis of the changes by using the theories mentioned above.

5.4.1   An Economic Explanation of Evolution of Entitlement System

In the changes of the ancient water rights structure of China, the entitlement system experienced the following three periods of changes. The Qin-Han period introduced the input quota entitlements, however, the government only announced some simple prohibitions and simple rules for water use and did not do much to develop it. The Tang-Song period introduced the resource quota entitlements in the allocation of water from different rivers, although input quota entitlement system remained dominant at the user level. Importantly, the government already formulated detailed rules. The Ming-Qing period introduced extensively the resource quota entitlement system in dividing up water use rights among farming households. The official institutions supplied by the state and folk rules were integrated organically to bring the entitlements system to a considerable height. Table 5.3 shows the changes of water entitlements systems in irrigation during the three different historical periods.

Table 5.3 Changes in the entitlement systems of irrigation water in ancient China

Period	Way of entitlements	Rules	Size of communities
Qin-Han	Input quota	Simple official rules	Whole irrigation area
Tang-Song	Input quota	Detailed official rules	Whole irrigation area, below tributary sluices
Ming-Qing	Resource quota	Developed official institutions and non-governmental institutions	Below tributary sluices Benefiting groups and households

The institutional change is a response to the change of environment constraints. From Qin-Han to Ming-Qing, the most significant change in the constraints on the option for entitlement systems was the increase in the value of irrigation water. In the Chinese history, the conflicts between humans and land were constantly aggravating. The population peak appeared in AD 2 in the Western Han Dynasty, when the population reached 59.59 million. The population peak in the Tang-Song period appeared in 756, when the population was 52.92 million; the population peak of the Ming Dynasty appeared in 1403, when the population was 66.6 million; the population at the beginning of the Qing Dynasty soared from 143.41 million in 1741 (the sixth year of the reign of Emperor Qian Long) to 432.16 million in 1851 (the first year of the reign of Emperor Xian Feng). Although the cultivated areas continued to increase, the average on the per capita basis dropped steadily. Before the Northern Song Dynasty, the cultivated land per capita averaged about 10 *mu* (1/15 a hectare). By the middle of the Qing Dynasty, the per capita cultivated land dropped to less than 4 *mu* (Gu, 1997). In this context, the value of irrigation water rose to a considerable level. The paddy fields had to pay more grain for water than arid crop land. During the reign of Emperor Qian Long in the Guanzhong Area, 'grain for water' accounted for about a quarter of the crop tax and the payment varied according to different grades of paddy fields. The principle was the volume of water used by burning incense; tax is paid on the use of water, which is divided up into three grades; when grain harvest increases, it attributes to water. So the levy is not heavy. It was exactly because of the rising value of water that a new institution was introduced to raise the efficiency of water use, hence providing external motivation for the introduction of resource quota entitlements system.

The institutional arrangement of resource quota entitlement system had to pay high costs in measurements and oversight while stimulating the intensive use of water resources. If the costs were too high, even the value of water was raised, the input quota should remain the most effective option. The resource quota entitlement system was introduced in history in that technical progress and institutional change helped change the costs of water measurement and oversight. In the Qin-Han period, the water allocation and measurement technology was very primitive. Till the Tang-Song period, such technology was developed, making it possible to realize water allocation in between different rivers and ditches. The Ming-Qing period used 'water path' to measure the volume of water and benefiting households used water path to measure water for irrigation. The technical progress helped lower the costs of water measurement. In oversight after water rights allocation, the costs were high when the state was directly involved. That was the case in the Tang-Song period. Up to the Ming-Qing period, the folk rules began to develop and non-official institutions began to play its role, thus greatly lowering the oversight costs in the enforcement of water rights. This is the result of institutional change.

In a word, the aggravating water scarcity from the Qin-Han to Ming-Qing made it possible for the introduction of institutions that were favorable for raising the efficiency of water use. On the other hand, technical progress and institutional changes helped lower the transaction costs in resource quota entitlement system. In the new institutional environment and technical conditions, resource quota became the institutional option for minimizing transaction costs. In addition, during the period of Qin-Han to Ming-Qing, the new institutional environment and technical conditions stimulated changes in the other aspects of the entitlement systems, including the increase in the acts of providing exclusive rights, clearer rules for the definition of property rights and ownership of resources by smaller communities as is shown in Table 5.3.

5.4.2   An Economic Explanation of Institutional Change of Initial Allocation of Water Rights

The most significant of institutional changes in the initial allocation of water rights for irrigation in ancient China was the evolution from license application in the Tang-Song period to water registration in the Ming-Qing period. It is recorded that in the old cases, application for water use must be filed before the sluice gates are opened, stating the benefiting households and what crops to irrigate. Sluice gates may be opened only after getting the license from the official in charge of water. The old cases here referred to Tang and Song dynasties. The Yuan Dynasty carried over such practice and it did not phase out until the Ming and Qing dynasties when a registration system was introduced. The water registration was a book recording water rights allocated according to the area of land. Once the registration was fixed, it was tantamount to local law, which was kept stable for a fairly long time to come (CIC, 2005).

The license application and registration is regarded as water allocation by administrative means, but they differ in nature. The water withdrawal license application was carried out every year and the water quotas changes with change in water resources and the types of crops irrigated. So the water rights quota was difficult to define. Water registration was, in essence, a certificate for the use of water authorized by the government. The water rights quotas were fixed unless the land rights changed. Under such system, water users needed not to declare the use of water with local government or water authorities. This system in fact enhanced the elements of law in water rights management. Compared with the pure administrative management, this system was obviously favorable for overcoming accidental and artificial factors and could reflect clearer the right relations (Wouters, Hu, Zhang, Tarlock, & Andrews-Speed, 2004).

Regarding the license application system evolved into water registration system, it is argued that under the technical conditions of the Tang-Song period, the license application system was an inefficient system, because the water measuring technology was not developed at the time, so did the water path method. Furthermore, Tang-Song period lacked the social environment on which water registration system depended, such as the degree of autonomy of non-governmental organizations and folk rules. The water registration system was not compatible with the then social environment and the costs would be extremely high. It should be admitted that the license application system was a rational option at the time. The water registration system became a reality only when the social setting changed, making it favorable for making such institutional arrangement that greatly lowered the costs for both static operation and dynamic changes.

The fundamental reason for the replacement of the license application system with the water registration system was that such a system lowered the costs for defining water rights. The license application system operated at a high cost and people of the Yuan Dynasty were already aware of it. It is pointed out that there were neither sequential orders of users in irrigating the land nor proper dates for irrigation. The water is limited to the size of land. If drought strikes, some people would use more water in violation of the limitation rules. If punishment is meted out, it would hurt the common people. If not, the egalitarian practice would never return. Those who are the first to irrigate would occupy more days. Those who get their land timely irrigated benefit while those who delay irrigation would miss the crop growth season (CIC, 2005; Gu, 1997). This shows that the license application system met a series of hitches. It did not only cost a lot in management on the part of the government but also achieved not so good a result in practice.

With the passing of time, driven by environment constraints, the government tended to seek a system that could economize on transaction costs. Till the Yuan Dynasty, the beginning form of water registration system appeared. A new way of water allocation is discussed that every water source has several sluice gates, which examine the number of water users. After the water is allocated, a date is fixed to open the sluice gates and all the water users will irrigate the land within 60 days. Then all the users together with their areas of farmland are recorded in a registrar. When the water path is measured from the upper reaches to down below, each sluice gate has a registration book and is fixed, showing who would use how much water and at what date. Thus, no one would dare to encroach upon the rights of others or violate the rules (Xiao, 1999). Obviously, such new institution was more convenient than the license application system, saving a lot of energy on the part of the government and the results were much better.

The institutional change from license application to water registration played an important role in the evolution of the water rights structure in ancient China. It was the precondition for the government in the Ming-Qing period to phase out and also an institutional change key to lower the costs of the operation of the water rights structure. This author's analysis shows that the institutional change was by no means accidental. It has profound endogenous rationale. In the Ming and Qing Dynasties when the contradictions between water and land were drastically worsening, the institutional change was inevitable as the water registration system greatly economized on transaction costs as motivated by the demand for raising the efficiency of irrigation systems and was therefore more attractive than the license application system. In line with technical progress and changes in social environment, the operational costs of the registration system per se greatly lowered the operational costs.

5.4.3 An Economic Explanation of Institutional Change of Water Rights Reallocation

The adjustment of the initial allocation of water rights for irrigation was undertaken by the government in the Tang-Song period when the license application system was in force. Annual water users applied for water use specified the size of their land and crops to be irrigated for the government to verify. This was in fact re-allocation of water rights by administrative means. When the registration system was introduced in the initial allocation, water rights trading began to appear, that is, water rights began to be reallocated by the market. In the discussion of Qingyu River and Longdong Canal, water rights trading cases are recorded in several irrigation areas (Chang, 2001, pp. 78–79):

> *According to the old rules, water was bought from the Yuancheng River together with the land and contracts were signed, specifying the terrain of the land and how the water flows. At the contract signing ceremony, river management official was invited to be present to measure the water path. If the management official was not invited, the contract was deemed as private dealings. The management official would establish that the seller was seeking profits while the buyer would be deprived of the right to use water. So there were water rules with the Longdong River while there were cases of land trading without water in the Muzhang River and there were cases of land trading with water path in the Yuancheng River. There were still trading of land without water path and there were people who did not invite local management officials. When local management official was invited to see the measurement of water path, the water would go with the land and buyer would have water rights. If local management official was not invited to see the water path, it was only land trading, without taking water path along. So the prices of land with water rights and those without were quite different.*

It shows that irrigation water rights trading was widespread and active in the Guanzhong area in the Qing Dynasty. In the previous dynasties, water rights trading was prohibited. Even in the late period of the Qing Dynasty when water rights trading was common occurrence, officials still denied that the government had the policy of allowing water rights trading. Water rights trading was not acceptable and it was deemed to bring conflicts among water users. However, it was unacceptable by the government, such trading really happened in local irrigation areas in the middle and late periods of the Qing Dynasty, at least in the Guanzhong area. This historical economic phenomenon merits attention and is of great value in academic studies. The following is a review of the explanations by some scholars before presenting the author's views.

It is believed that the separation of irrigation water rights from the land use rights marks the starting point of water rights entering the market in the Guanzhong area during the Qing Dynasty and also the preconditions for the lot of water rights trading to happen (Xiao, 1999). Xiao examines in detail the process of water rights alienating from land rights, pointing out that it is an inevitable results of the historical development and lists two factors contributing to the separation of water rights and land rights: the strict correspondence of water rights to land rights does not conform to the efficiency principle for the operation of agricultural technology and the rising value of irrigation water stimulate the process. Further, water rights trading in the Ming-Qing period is presented in these aspects (Chang, 2001): (1) the spread of commodity economy awareness; (2) development of the commodity economy; (3) drastic increase in population, which pushed up the relative prices of land and water resources; (4) the rise in the irrigation water value, which, in turn, gave rise to institutional innovation, ultimately leading to the institutional change in the market allocation of water resources. Since the Qin-Han period, the value of irrigation water assumed an upward trend and that gave a bigger push to water right trading. However, the real problem lies in why demand for market in the reallocation of water rights happened in the Qing Dynasty but not earlier, due to the relevant price rise in water resources.

Under the license application system, the government annually allocated water rights for irrigation by administrative means and the exclusiveness of the water rights were very limited and the quantity of water could not last due to frequent changes. With the strict tally of water rights with the size of farmland, the water rights quota was not divisible, let alone transferable. It can be understood that the water rights quality θ under the license application system was very low, close to zero in the Tang-Song period, compatible with the allocation of water rights directly by the administrative means. When the license application system evolved into water registration system, it granted farming households with relatively stable and lasting rights. Under the water rights quota constraints, water may be allocated according to different sizes of land, thus endowing the water rights with divisibility. Due to the rise in exclusiveness of water rights, such rights tended to become the private property of the owners who can dispose and transfer them freely. That means that the water registration system can greatly raise the water rights quality θ. The longer the allocation system lasts, the closer to 1 the θ value tends to be. This process is reflected in Fig. 5.5. The figure shows perceptually that after the evolution over the thousand years, the water rights quality of the Ming and Qing dynasties was much higher than in the Tang and Song dynasties.

The changes in the initial allocation mechanism brought about changes in the irrigation water rights quality. This had two-pronged effects. The first is that the cost for adjusting water rights by administrative means rose; and the second is that the costs of market reallocation were lowered. As the administrative means and market means were mutually exclusive, the two effects mentioned above were in reality the two sides of the same coin. Other institutional arrangements were compatible with the water registration, such as fixed time irrigation and folk rules, which further lowered the cost of reallocation of water rights by market. It is just because the changes from the license application system to water registration system that the water rights quality improved. The effective changes in the relative costs of water rights allocation led to the active trading of water rights in the middle and late periods of the Qing Dynasty. This is a typical process of change that induced the institutional change. Driven by the inherent force of lowering transaction costs, market means had greater advantage with regard to transaction costs in the adjustment of water rights. This explains the replacement of the administrative means by market means.

Water rights trading was always a kind of non-governmental act in ancient China and it had no legal status recognized by the state. This, in fact, reflects the difficulty behind water rights trading. The introduction of the market force was, on the one hand, the inevitable result of social and economic development and on the other hand seriously disrupted the traditional social order, thus making water rights trading not only a simple economic issue but an issue with heavy political colors. The main function of the market is to seek efficiency. Water rights trading is a response to the rise in the relative prices of water resources. Its main function is to raise the efficiency of resources allocation. However, the allocation of water rights by administrative means seeks equality that is more of a political objective. We can then understand why the governments in ancient society held the negative view to water rights trading. For the government, the favorable factor concomitant of water right trading is a force detrimental to the traditional order.

We can further deduct that in the water rights re-allocation process in the Ming-Qing period, administrative and market means existed alternatively. The administrative means always made the initial water rights allocation according to the principle of water rights corresponding to land rights. After it completed its functions of achieving equality, the administrative means would phase out to give way to market means, driven by the demand for raising efficiency, which play the role of adjusting the initial allocation of water rights. With the large amount of water rights trading, water rights gradually alienated itself from land rights and the allocation of water rights more and more alienated from the equality principle until it became so intolerable in the society that administrative means would come back again to make re-allocation. So the social order would return to square one, thus completing a round of replacement in the reallocation of water rights. What the market does is to adjust the initial allocation of water rights within a small scope while the administrative means changes the water rights on a larger scale. In the serpentine course of conflicts, the mutually exclusive administrative and market means has become coupled. This is testified by the following historical materials:

According to a record on a stone tablet (1588 AD) in Jiexiu County in North China's Shanxi Province, irrigation development started in the county in the Northern Song dynasty. In previous dynasties, water was measured according to the size of farmland and water was used by turns, and there were water headmen and river management officials responsible for the matter. But with the passing of time, the system revealed its defects when seizure by force took place. In the 25th year of Emperor Jia Qing of the Ming Dynasty (1545 AD), county magistrate Wu Shaowei intervened and restored the former practice. Then trading of water took place. In the first year of the reign of Emperor Long Qing (1567 AD), another county magistrate Liu Pang changed the water path in force and introduced a new method. Every village had to have a registration book. Although it eased the conflicts for the time being, there was still land without water and water without land and water and land could be traded separately. In the 15th year of the reign of Emperor Wan Li (1587 AD), county magistrate Wang Yikui remeasured the land to distribute water rights, with every family having a registration book. At the same time, he ruled that the trading of land and water may proceed together (Gu, 1997; Xiao, 1999). The drawback of water trading reflected in the tablet refers to the fact that the acquisition of water rights enabled rich people to get water for their dry land without paying taxes while the poor people had no water to use although they paid taxes for water and land, which led to social instability. However, people at the time only saw the defects of water trading, without recognizing the advantages brought about by the market allocation of resources. That is the root cause economically for the rampant water trading in disregard of the government ban. It is exactly these disadvantages and advantages that led to the cyclic major adjustments of irrigation water rights that lasted for more than 20 years (Xiao, 1999).

According to historical materials available about Yuandeng River in Guanzhong toward the end of the Qing Dynasty, water registration book was renewed in the 16th year of the reign of Emperor Qian Long, in the years of Emperor Jia Qing and the 22nd year of the reign of Emperor Dao Guang. Each renewal verified the size of farmland. If it was found that land rights were transferred, the name of the beneficiaries would be changed in order to equalize water rights. But due to rampant water rights trading, the size of farmland and water rights quotas failed to tally. It only played the role of checking when each village started or terminated the use water. The example seems to show that in the middle and late periods of the Qing Dynasty, the cycle for major adjustments of water rights by administrative means were greatly extended and the pattern of water rights allocation by water trading got confirmation and inheritance. We may deduce from such change that in the middle and late periods of the Qing Dynasty, market adjustments of water rights for irrigation operated well and played a bigger role than administrative means (Twitchett & Fairbank, 1978; Zhang, 2012).

The prevalence of water rights trading proved that more and more market elements were introduced into the governance structure and indicated that the degree of hierarchy of the governance structure was lowered. In other words, after the evolution to the last empire of ancient China, there was no longer a pure hierarchy structure in both water governance and state governance. This is a natural evolution aimed at lowering management costs in the long-term operation of the hierarchy structure. In order to economize on management costs, the state delegated the micro-management power to non-governmental agents. The lowering of the degree of hierarchical governance structure may is considered as the price paid by the state. Just as the liberalization of water rights trading is very limited, no matter how limited the drop in the degree of hierarchy in state governance structure is. On the whole, all the empires in ancient China were alarmingly able to maintain the unique hierarchy governance structure.

Chapter 6
Water Rights Structure and Economic Explanation: Empirical Study of the Yellow River Basin in Modern China

This chapter studies the changes of the surface water rights structure of the Yellow River basin since the founding of New China. First, it examines the basic characteristics of such structure during the planned economy period, then the process of change during the reform and opening up period and then the characteristics of the current structure. Then it devotes the study to the changes in the irrigation water rights allocation mechanisms in two periods before going on to the rational explanation of the changes of the water rights structure since the founding of New China, with emphasis put on the entitlement system, including (1) entitlement system in the irrigation water rights allocation since the country introduced reform and opening up policies, when resource quota was extensively replaced by input quota; (2) the differences in the entitlement systems in the irrigation areas of the upper reaches and the lower reaches and their differences in path option in institutional reform; (3) resource quota entitlement system widespread at the other levels of water rights allocation. As the current water rights are allocated by administrative means, it conforms to the hierarchical water rights structure studied in the previous chapters and so there is no need for repeated explanations. This chapter dedicates a considerable space to examining the operation of the current water rights structure and the relations between the current water resource management system and the quality of water rights. By analyzing the data at the three levels of the Yellow River basin, local governments and groups, this chapter carries out effective testing of the current water rights allocation by administrative means and sum up the theoretical implications of the water allocation practice since 1987 in the Yellow River basin (Xia & Pahl-Wostl, 2012).

© Springer Nature Singapore Pte Ltd. 2018
Y. Wang, *Assessing Water Rights in China*, Water Resources Development
and Management, DOI 10.1007/978-981-10-5083-1_6

6.1 Water Rights Structure in the Planned Economy Period (1949–1977)

6.1.1 Background Information

Since the founding of the P.R. China, irrigation in the Yellow River basin has developed rapidly, with the irrigated areas expanding by five-fold the areas before the country introduced reform and opening up policies. Starting from scratch, the Yellow River water irrigation areas in the lower reaches had expanded to more than 1,333,333.3 hectares up to the country's reform and opening up. During the whole planned economy period, a great drive was carried out to build irrigation projects in the river basin, with the water volume used assuming a rapid upward trend until it was four times as much as that used in the post-liberation days (see 5.1).

In 1950, the central people's government issued instructions to change the Yellow River Conservancy Commission, a joint Yellow River Control organization formed by the three provinces of Shandong, Pingyuan and Henan, into the Yellow River Basin Water Committee to unify the development and control of the whole Yellow River basin. But in reality, during the whole of the planned economy period, the Committee focused its work on ensuring the security against floods in the lower reaches, failing to display its water management functions. The Yellow River water resources were subject to the management separately by various provinces and regions.

In 1954, the State Planning Commission mapped out an integrated water utilization plan for the Yellow River basin to allocate, for the first time, the water resources of the entire river. According to the plan, the anticipated long-term demand of Yellow River run-off for irrigation was 47 billion cubic meters and allocated the volume to riparian province and region. The 1954 plan embodied the idea of exhausting the water of the Yellow River, leaving little for the ecology. As there was no mechanism to execute the plan, the plan, in the end, was pigeonholed (Yahua Wang, 2013a).

From the 1950s, the water extracted from the Yellow River increased greatly, with the demand by some provinces and regions being unable to be satisfied during key water using seasons. In sections of the upper and lower reaches where water use was concentrated, provinces and regions concerned agreed upon a proportionate allocation of the water through consultation. In 1959–1961, Henan, Shandong and Hebei on the lower reaches concluded an agreement on diverting water during dry seasons according to the ratio of 2:2:1 to Henan, Shandong and Hebei from Qinchang (about the place at Huayuankou at present). In 1962, the Yellow River water diversion stopped and no more allocation was made after irrigation resumed. Ningxia and Inner Mongolia on the upper reaches agreed to divert the water according to the ratio of 4:6 since 1961, which lasted until the 1990s when the YRCC unified the regulation of the water volume in the whole river basin. Before that, the utilization of water resources between the upper and lower reaches was not mutually affected (Yahua Wang, 2013a; Y. Wang and Tian, 2010).

The coordinated allocation of water among different provinces and regions was only temporary measures when the water resources were not enough and it lacked the mechanism for effective enforcement. There was no official water allocation mechanism among different provinces and regions. In the whole of the planned economy period, the water affairs administrations from the central government down to grassroots were very weak and there was no resource management at all within the river basin. The Yellow River water resources were in a state of segmentation and disorder, a practice described as those who develop the resources use and own them.

Small irrigation facilities built by individuals and mutual aid teams were gradually taken over by cooperatives after the founding of New China, subjecting to the unified management by cooperatives. Large and midsize irrigation projects that benefited different townships, counties or even regions were subject to management by special organizations that worked under the leadership of departments at superior levels. Special organizations and grassroots water use units were subject to management by integrating specialized organizations and the local people (RIDMWR, 1999). Acted by the institutional environment, there was a tendency of stressing construction to the neglect of management; stressing backbone projects to the neglect of supporting projects; and stressing engineering to the neglect of practical results in the management of irrigation areas.

6.1.2   Characteristics of Water Rights Structure in Planned Economy Period

The surface water rights structure of the Yellow River basin during the period of planned economy is shown in Fig. 6.1. Compared with the ancient China, it has the following characteristics:

First, with the volume of water captured, the number of policymaking entities at the group and use levels increased drastically. Statistics show that there were about 30 irrigation areas each covering an area of more than 666.6 hectares before the founding of New China. However, the number soared to 671 by 1985 and the riparian population increased from 37 million in 1949 to 81.53 million by 1980. That means that the number of policymaking entities in the then water rights structure shot up rapidly.

Second, it appears that basin water organizations took charge of development and water control in the whole basin as management agents commissioned by the central government. In the period of planned economy, the basin organizations did not play their substantial role in water resource allocation and management. Rather, they acted as technical advisers only to the water resource department of the central government.

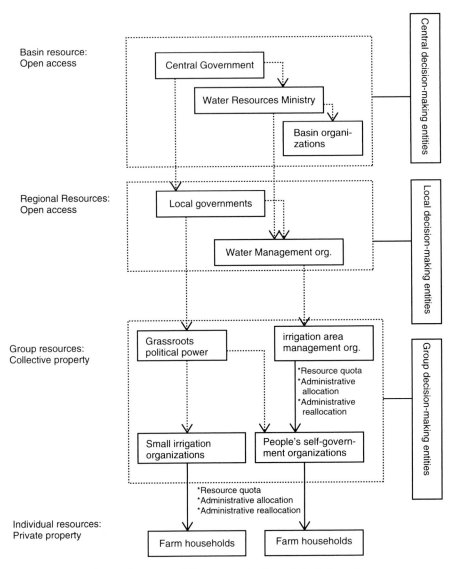

Fig. 6.1 Yellow River surface water rights structure in the period of planned economy

Third, with the rise in the water scarcity in the basin, necessity arose in the allocation of water resources between localities and between groups. In the period of planned economy, the contradictions among different localities and groups were mainly settled through artificial intervention. There was no institutionalized mechanism of water allocation. The water resources at the basin and local government levels were in a state of quasi-open access while they were divided up at the group level as the collective property.

Fourth, the water allocation from the group level to user level, namely the allocation of water for irrigation, was subject to centralized and unified allocation and management at different levels. The initial allocation and re-allocation were mainly done by administrative means. In irrigation areas where water resources were better managed, the practice of contracting for flow volume or water volume at different sections of the river and water volumes were divided up between weirs and tributaries by way of resource quota. Water for use in irrigation experiments, leveling land and rebuilding ditches and low-lying land were also brought under management. In general, most irrigation areas had ineffective water allocation mechanism at the user level, where the same water resources were shared according to the egalitarian principle.

6.2 Water Rights Structure in the Transition to Market Economy (1978–2002)

6.2.1 Background Information

Since the Reform and Opening-Up Policy, China has changed the general environment and swiftly established an initial water resource management system. The past over 20 years marked a big development in water management system, with allocation mechanism established at all levels of the hierarchy and a number of institutions introduced, and changes at a speed that has never been seen in history. We divide this period into three stages according to the changes in water rights structure.

The first stage covers from 1978 to 1986, which was a period of preparations for changes. Some sections of the water course in the lower reaches of the Yellow River began to dry up starting from the 1970s, indicating that water resource scarcity was expanded to the whole of the river basin, demanding unified allocation of the whole river water. Starting from 1982, the State Planning Commission set out to plan the utilization of the water resources of the river and called related departments of various provinces and regions to discuss matters concerning water allocation so as to prepare a water allocation scheme. Some provinces and regions in the basin strengthened institutional arrangements for water allocation. Shanxi, for instance, introduced the water use permit system in its local water resource law in 1982 and started to levy water resources fees. This raised the planned utilization of the water resources in all the irrigation areas, which started water fee reform. During this period, water rights allocation was expanded from projects to the whole river basin and part of the areas started water allocation mechanism between local governments and groups, resulting in the reform of the irrigation management system and laying the foundation for the change in the water rights structure.

The second stage covers from 1987 to 1997, when water rights system developed rapidly. In September 1987, the State Council approved the scheme of water supply in the Yellow River basin and issued it to 11 provinces and regions and related State

Council departments, thus starting the allocation of 37 billion m³ of waters among the riparian provinces and regions (see Table 6.1). The 1987 water allocation scheme marked the beginning of water allocation among the riparian provinces and regions and served as the basis for water rights allocation among different provinces and regions. In 1988, the *Water Law of the People's Republic of China* was promulgated, officially fixing the basic systems on the management of water resources. According to the provisions about water permit system in the Water Law, the State Council issued the detailed rules for the implementation of the water extraction permit system in 1993, specifying the scope, application, examination and approval, issue, invalidation and permit revoking procedures. In 1994, the Ministry of Water Resources authorized the YRCC to exercise full-quota or limited quota management of the water extraction permits for the main course and tributaries spanning provinces and regions and exercise total control for water extraction according to the water volume available for allocation. The YRCC then demarcated the terms of reference for extracting water for basin organizations and various provinces and regions and exercised control. The water extraction registration was completed in the first half of 1996 and then shifted the emphasis of its work to water permit oversight and management. After the detailed rules for the implementation of the water permit system was issued in all provinces and regions, various provinces and regions also began to strengthen water extraction management within their terms of reference. This new framework for the management of Yellow River water took shape, featuring the integration of management at the level of basin organizations and administrative regions. This period also witnessed big progress in the collection of water fees and levies on water resources. In a word, through the ten years of institutional reform, water allocation mechanisms were introduced between local governments and among water extraction groups, thus giving rise to a complete framework of water rights structure.

The third state spans from 1998 to 2002, a period of improvements of the water rights system. The second stage left a number of defects in the institutional designs and enforcement, including irrationality in the definition of initial rights, imperfection of the water permit management, especially the enforcement mechanism which was not put in place, thus making the scheme difficult to carry out. Some province extracted more water than allocated to them; some water disputes were not settled well. That is an important reason for the dry-up of the river sources in some sections in the 1990s. The continuous water shortages and river course dry-up aroused full attention from the central government. Starting from 1998, within a very short period of time, a large number of new measures were introduced, especially mechanisms for enforcing the allocation of water rights. In 1998, the Ministry of Water Resources and the State Planning Commission jointly issued a scheme for the annual water available for allocation and the regulation of water volume of the main river course. In 1999, the YRCC began to exercise the unified regulation. For this purpose, the Committee set up a department specializing in the water regulation. This monitoring and control mechanism with administrative means as the principal ways was enhanced and the hydrological monitoring was also strengthened. The Water Law, which was revised in 2002, was promulgated, thus codifying

Table 6.1 Water allocation scheme before the south-to-north water diversion project took effect

Place	Qinghai	Sichuan	Gansu	Ningxia	Inner Mongolia	Shaanxi	Shanxi	Henan	Shandong	Hebei & Tianjin	Total
Annual water consumption (million m³)	1410	40	3040	4000	5860	3800	4310	5540	7000	2000	37,000

the water management experience over the past dozens of years. It provides that water resources are owned by the state and the ownership of water resources is exercised by the State Council on behalf of the state. It also specified the compensatory water use system, and water management system that integrates basic management with administrative regional management. The new water law also established a series of water management systems. It, in fact, established a complete water rights allocation mechanism, including how to allocate water, who have the rights to use water and how to use it, covering all aspects of water rights allocation. The new water law officially established the legal status of basin management organizations, granting them with the powers of unifying the management and regulation of water in the Yellow River basin. Under the guidance of the new water law, a series of important laws and regulations are being drafted, such as regulations on the management of Yellow River Water Resources and the Yellow River Water Law. These are expected to play a big role in improving the basin water resource management system.

6.2.2 De Facto Water Rights Structure

The years from 1988 to the present is a period of rapid development in the institutional arrangements concerning water resources and also a period of big development in the Yellow River water rights. At present, a complete hierarchical structural framework has taken shape, giving rise to a complete set of planned allocation and use of water from the riparian provinces and regions down to various water extraction stations and down to water users. Compared with the period of planned economy, the more than 20 years of reform have brought about tremendous changes in the current water rights structure.

First, the basin organizations representing the central policymaking entities have enhanced their management functions. From the period of planned economy to the mid-1990s, China's water resource development and utilization were subject to management at different levels, resulting in segmentation. In 1998, the State Council clarified that the Ministry of Water Resources is the State Council department taking charge of water affairs and responsible for unifying the management of all water resources in the country. After that, the Ministry of Water Resources approved the schemes mapped out by the YRCC, clarifying that the Committee represents the Ministry of Water Resources in exercising the powers of administration in all river basins. In 2002, the new water law further clarified the legal status of basin management organizations, stipulating that basin management organizations are organizations set up by State Council departments in charge of water administration in major rivers and lake areas to enforce the law and administrative decrees within their jurisdiction and water resource management and oversight as authorized by the State Council department in charge of water administration.

Second, the terms of reference of local policy making entities and central policy making entities were clarified. According to the spirit of the reform of state organs in 1998, it got clear about the State Council department in charge of water administration and also about the water administrations at the level of local governments. These are the water resources departments of various provinces and regions and water resources bureaus of various cities and counties. The 2002 New Water Law established the integrated system of basin management and regional management and all local governments in the riparian areas are required to exercise water resource management according to their respective terms of reference. While holding on to the principle of unified management of Yellow River water resources, the YRCC, in the capacity of basin management organization, represents the state to exercise basin management, enforcing the legal provisions and the management and oversight power authorized by the Ministry of Water Resources. Water resources administrations at local levels act as regional management representatives to exercise water resource management within their respective administrative areas according to law.

Third, complete water allocation mechanisms were established between riparian provinces and regions. The scheme of water available for supply issued by the State Council in 1987 was, in fact, the definition of initial water rights for riparian provinces and regions. According to the quotas for water available for allocation in normal years and the proportions of the year's water available for supply, the 1998 plan for annual allocation of water available in the Yellow River and the water volume regulation in the main river course and the rules on the Yellow River water regulation, issued by the Ministry of Water Resources and the State Planning Commission, fixed the annual quotas for various provinces and regions, making the monthly quotas divided in the same proportion. The annual quotas allocated to various provinces and regions were subject to total volume control and provided for the water allocation in special dry years. This series of institutional arrangements are, in fact, a complete set of water rights allocation mechanisms between the riparian provinces and regions.

Fourth, a complete rights attributes management system was established at the group level, to be implemented by basin organizations and local water administration departments according to their division of labor. In May 1994, the Ministry of Water Resources issued a circular, granting the rights to the YRCC the power of approving and issuing water permits above the quotas. The YRCC then formulated detailed rules on the implementation of water permits regulations. Till the end of 1999, the Committee issued 323 water extraction permits, with a total water volume of 30.3 billion m^3 (Yahua Wang, 2013b). The riparian provinces and regions, according to their terms of reference, carried out water extraction permits registration and management within their respective jurisdiction. The implementation of the water extraction permit system in the whole Yellow River basin stimulated the establishment of planned water allocation mechanism between different water extraction groups, with their respective rights defined.

Figure 6.2 shows the surface water rights structure in the upper and middle reaches of the Yellow River and Fig. 6.3 shows such structure in the Yellow River irrigation areas in the lower reaches. The figures show that a planned water use system has been established in the whole of the Yellow River basin, with rights

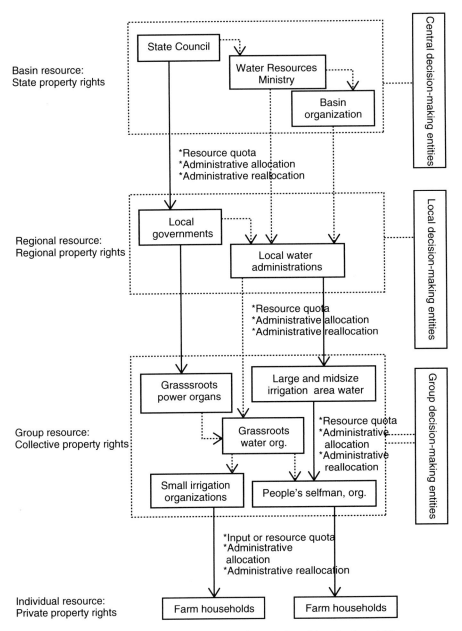

Fig. 6.2 Current surface water rights hierarchy of the Yellow River in the middle and upper reaches

divided up among all levels. The water rights allocation pattern at the basin level is the state unifies and controls the allocation of water volume, provinces and regions are responsible for specific water allocation and unified regulation of major water

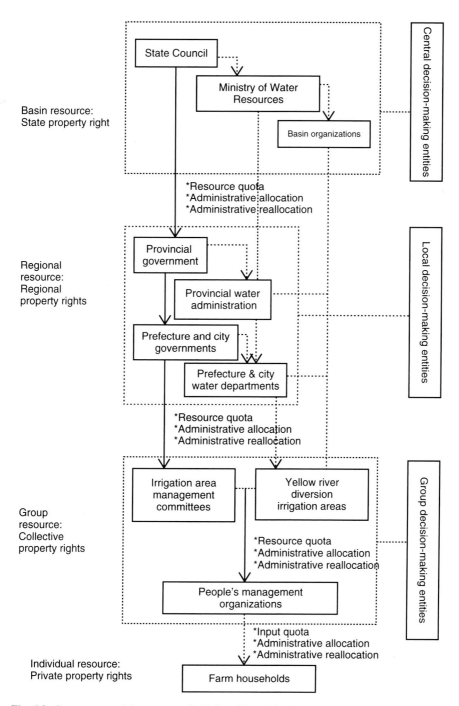

Fig. 6.3 Current water rights structure in Yellow River irrigation area in the lower reaches

extraction outlets and backbone reservoirs (Yahua Wang, 2013a). Provinces and regions also established their own systems for planned use of water. Some provinces and regions have allocated water volume to different areas. Most of the irrigation areas have been covered by the water extraction permit system.

The figures also reveal the disparities in the water resource management between the lower and middle and upper reaches of the river. Shandong and Henan in the lower reaches of the river have set up water administrations at all levels, including province, prefecture and county bureaus and grassroots water administrative sections, which exercise permit management in the lower reaches. All the culverts and weirs are subject to the control and management by the basin organizations. There are also management bureaus in the upper and middle reaches to control the main river course and manage the water extraction from big water extraction outlets. But the weirs in the upper and middle reaches of the river are subject to the management by localities. The differences in the management system have endowed the basin organizations greater power in the lower reaches than in the upper reaches, making it possible for the organizations to get involved in the direct water allocation at the group level while achieving the same purpose by controlling the water use volume by provinces and regions in the upper and middle reaches, the water use and allocation are under responsibility of water administrations of provinces and regions.

6.3 Changes of Water Rights Allocation Mechanism Since the Founding of New China

Irrigation water rights allocation is the main subject in the study of the changes of ancient water rights structure. This is associated with the single nature of water used in the ancient society. Since the founding of the P.R. China, the water resource utilization has been diversified, with industrial and city use rising rapidly and water allocation extending to all the riparian provinces and regions in the whole river basins, which was rarely seen in the ancient times. Despite all these, agriculture remains the main water user in the Yellow River basin. This section gives a brief account of the changes in the irrigation water management in the upper and lower reaches of the Yellow River.

6.3.1 Irrigation Water Rights Allocation in Planned Economy Period

Irrigation in the Yellow River basin exhibited some differences in the upper and middle reaches and the lower reaches after the founding of the P.R. China. The increase in the areas of irrigated land with water diverted from the Yellow River in the upper and middle reaches of the river was the result of large scaled expansion and rebuilding of irrigation projects. Most of the localities have a long tradition of irrigation and therefore the irrigated areas had fairly complete systems for planned use of water before the country introduced the reform and opening up policies. In the irrigated areas using water diverted from the Yellow River in the lower reached started anew after the founding of the P.R. China. During the period of planned economy, main efforts were focused on project construction, water management being downplayed, and a complete system of planned allocation of water had not been developed.

In the upper and middle reaches of the Yellow River, most irrigation areas have had an irrigation tradition of about 1000 years. With the changes in social structure and technical progress, the traditional management system was gradually brought onto the orbit of planned water allocation. Irrigation technologies in the modern sense was introduced from the nationalist period and gained great ground after the founding of New China. For new canals, flow meters and other water measuring instruments, such as level gauge, level meter, water measuring bank, flow meter, buoy and water meter were introduced. The coverage of hydrological stations has been expanded. With the popularization and spread of the flow meter, the water allocation according to a fixed area of land and contracting for a fixed amount of water has become a reality, giving rise to a new system of planned use of water at all levels. This period witnessed the shape-up of the system of planned use of water in part of the upper and middle reaches, with complete organizations and water allocation and use system established.

The history of irrigation in the Yellow River bend area of Inner Mongolia and the Qingtong Gorge area of Ningxia was dated back to the Qin and Han periods. The areas serve as excellent examples in the characteristics of water rights allocation in the irrigation areas of the upper and middle reaches of the river during the period of planned economy. The Yellow River bend area is the largest irrigation areas with water diverted from the Yellow River. Although organizations were restructured in the post-liberation days, the area follows the practice before the founding of the P.R. China in the management of water allocation and use. From 1959 to 1961, a key water project was built at Sanshenggong on the main river course to replace the practice of diverting water from multiple points without dams, with water volume going beyond control. The areas thus began to introduce planned use of water. Plans were mapped out for water quotas for different crops and water was diverted and allocated at different levels in a planned way. Then, the irrigation management system underwent a series of changes that have brought the planned water use system to near perfection. The Qingtong Gorge irrigation area in Ningxia is the second largest irrigation areas with water diverted from the Yellow River. Similar to the Yellow River Bend area, Qingtong Gorge irrigation areas also follows the old practice of water allocation at the beginning of the founding of New China. However, the area began to strengthen water management in the 1950s when there were more frequent dry season flows. After 1960, the area is introduced flow meters and formulated plan for allocating water according to irrigation areas, crop types, the volume of water needed and time. Through more than 20 years of development, a complete system of planned use of water took shape in the 1980s (Yahua Wang, 2013a).

Experiments in diverting water from the Yellow River for irrigation was proved a success in the lower reaches of the Yellow River at the beginning of the 1950s. Irrigation has developed by leaps and bounds since then. But flooded irrigation to the neglect of water drainage caused a large expanse of land to be salinized and irrigation was forced to stop in 1962. During the years of the period of planned economy, a large scaled water diversion projects were carried out, mostly carried by local governments. But the efforts were mainly focused on irrigation and drainage and field supporting projects, water management conceded to the secondary place, resulting in an extensive management in the whole irrigation area. Although water management was strengthened after irrigation resumed, the water allocation mechanisms was incomplete during this period. After a long period of exploration, the irrigation was gradually brought onto the path of rational water diversion from the Yellow River and scientific irrigation after the 1980s.

6.3.2   Irrigation Water Rights Allocation in the Period of Transitional Economy

Since reform and opening up, the Yellow River basin have all strengthened management of the irrigated areas and established complete organizational setup and water use management systems. Most irrigation areas organizationally combined specialized management with people participation. The people's participation in management in the upper and middle reaches has developed fuller. However, the irrigation areas in the lower reaches of the Yellow River depends more on specialized management and the level of people's participation in management is relatively low. In management, all irrigation areas have established the system of planned water use, which has been raised in its scientific nature and in the rationale of water allocation thanks to technical progress. In terms of water price, the prices of water supplied by water projects have been brought into the system for commodity price management of the state and the pricing, approval and collection have been standardized. The prices have been adjusted on a number of occasions. A pricing mechanism with added prices for extra-quota water used and two-part water pricing has taken shape.

First of all, in terms of organizational setup, all the large and medium-sized irrigation areas in the Yellow River basin have established a system that integrates specialized management with popular management (Y. Wang and Tian, 2010). Take the Ningxia and Inner Monglia irrigation areas. Specialized management organizations are responsible for public managed channels, including management, maintenance, servicing, water transport and allocation and flow measuring of trunk canals, branch canals and general drainage canals, dry ditches, branch ditches and publicly managed straight channels for water allocation. The people's management organizations are divided into two levels: water management stations at the township level and branch sluice gate management committees or farmer water users association, all subject to the leadership by water resource bureaus at banner (county, city) and township governments. All the township in the irrigation areas have water control stations, being responsible for the building and management of irrigation and drainage projects and water transport and allocation. All the branches canals and channels with weirs committees have members and chiefs. All villages have full-time water controllers. All villages have household-based irrigation contract groups to responsible for the drudge of channels, project maintenance, water transport and allocation and water rate collection in their own villages or within the scope in their respective jurisdiction. Part of the places in the irrigation areas has established water user associations, with the owners of all beneficial farm households being members. All water using groups (or villagers group) elect representatives, which will form the executive board of the associations. The associations are mainly responsible for the project management, including both irrigation and drainage, flow measure and water rate collection within the coverage of their canals (CIC, 2005).

Now we discuss about the water use system. As early as in 1980, the Shaanxi provincial government issued the "*Regulations on Water Resources Management for Trial Implementation*", which provided that all irrigation areas have to use water in a planned way; water rights must be centralized and water must be regulated in a unified manner and managed at different levels. No unit or individual is allowed to regulate water at will. Inner Mongolia, Ningxia, Gansu and Shanxi in the upper and middle reaches have issued similar regulations. The irrigation areas using water diverted from the Yellow River began to strengthen water management starting from the beginning of the 1980s. In 1994, the Ministry of Water Resources issued the "*Regulations on the Management of Irrigation Using Water Diverted from the Yellow River in the Lower Reaches*", which clearly defined the internal water allocation system. Water shall be supply according to water tickets. All irrigation areas must, according to their project conditions, formulate water allocation and use systems for implementation after discussion by representatives from all irrigation areas. The irrigation management units shall unify regulation of water volume and delegate management power to various levels (Zhu, 2004). In general, most of the large and medium-sized irrigation areas realized flow and volume contracting. In recent years, part of the irrigation areas has carried out experiments in establishing water users association and the reform of contracting for branch canals, thus stimulating the rise in the management level and economizing of water at the grassroots level. There are also parts of the areas that have introduced reforms of the property rights of rural small irrigation projects to explore such ways as contracting, water users association, auctioning, leasing and joint stock systems in the reform of the water property rights system. For instance, Qingtongxia City in Ningxia started the reform of property rights of small irrigation projects in 2002. The whole city formed 25 farmer water user association in a short space of two years, contracting for 209 branch canals. The Areas covered by property right reform accounted for 89.6% of the total irrigated areas and the number of channels and canals contracted accounted for more than 935. More than 53.44 million cubic meters of water were saved over the past two years and farmer users paid 900,000 yuan of water fees (X. Chen, 2002; Pietz, 2010; Yahua Wang & Hu, 2002).

Lastly, let us look at the reform of the water prices. First, let us come to the water prices at the head of canals in the lower reaches. In 1980, the head of the canals in the lower reaches of the Yellow River began to collect water fees from units that diverted water for irrigation from the Yellow river. In 1982, the Ministry of Water Resources and Power issued the rules on collecting water fees in the low reaches of the Yellow River. In 1989, the Ministry of Water Resources issued the collection and management of water fees from the projects diverting Yellow River Water for irrigation. In 1990–2000, the prices at the head of canals remained unadjusted. In 2000, the State Planning commission issued a circular on adjusting the water prices for projects diverting Yellow River water for irrigation. Although the adjustment range was big, the current prices were still hardly ½ of the cost. Let us look at the Yellow River Bend Irrigation Area in Inner Mongolia. Before 1980, water prices were paid in grain according to the areas of crops. In 1981, reform was carried out and water rates began to be collected at the head of canals and different prices were fixed in different sections. In 1988, further reform was carried out and water rates were collected on the per cubic meter basis at flow measuring points. In 1995, the irrigation areas began to introduce the reform according to the principle of running at double small steps to raise water rates every year by a small margin. In 1999, the area began to introduce the policy of *"planned use of water, fixing prices according to different sections of canals and imposing addition prices on water in access of planned amount"*. Although the prices were adjusted time and again, they still remained at about 75% of the cost by 2001. The same was true with the irrigation areas in Ningxia. Although water prices were adjusted many times over the past 20 years, the price still accounted for 47.4% of the water supply cost (Yahua Wang & Hu, 2002; H. Yang & Jia, 2008). The changes in the water prices in the Yellow River Bend Area since the founding of new China are shown in Table 6.2.

6.4 Economic Explanation of Changes in the Yellow River Surface Water Entitlement System Since the Founding of New China

6.4.1 Economic Explanation of Changes in the Irrigation Water Entitlement System in the Upper and Middle Reaches of the Yellow River

This author has selected the Yellow River Bend Irrigation Area in Inner Mongolia as a paradigm in the upper and middle reaches of the Yellow River for the study of the changes in the irrigation water entitlement system. Table 6.2 shows the process

Table 6.2 Changes in the water prices of the inner Mongolia Yellow River bend irrigation area in 1949–2001

Year	Mean rate	Of which		Note
		Price for summer	Price for Autumn	
1949	2.5 kg of millet per mu			Before liberation
1952	2.5 kg of millet per mu			3% of average per mu output
1958	4.2 kg of millet per mu			5% of the average per mu output
1959	2.5 kg of millet per mu			3% of the average per mu output
1964	3.35 kg of millet per mu			4% of the average per mu output
1980	4 kg of millet per mu			
1981	0.114 cents/m^3	0.1	1.5	Water volume at outlets of trunk canals
1984	0.18 cents/m^3	0.16	0.23	Water volume at outlets of trunk canals
1988	0.6 cents/m^3	0.46	0.92	Water volume at weirs outlets
1989	0.9 分/m^3	0.8	1.2	Water volume at outlets of weirs
1995	1.7 分/m^3	1.65	2.1	Water volume at outlets of weirs
1996	2.0 分/m^3	1.9	2.4	Water volume at outlets of weirs
1997	2.3 分/m^3	2.2	2.7	Water volume at outlets of weirs
1998	3.3 分/m^3	3.1	3.8	Water volume at outlets of weirs
1999–2001	4.0 分/m^3	3.8	4.7	Water volume at outlets of weirs

Source: Inner Mongolia Yellow River Bend Irrigation Area management Bureau, "History of Irrigation in Bayannur League", 2002, p.352

of gradual shrinking of the equitable sharing of water fees, which, in fact, reflects the change of the entitlement system from input quota to resource quota and the trend of the shrinking of the size of communities still following the resource quota. Before 1980, the whole irrigation area paid water rates in grain on the per mu basis, which is an input quota entitlement system. Starting from 1981, water rates began to be collected on the cubic meter basis, thus introducing the resource quota system among the trunk canals and irrigation outlets. The method was further spread to between weirs. In 2000, part of the irrigation area began to charge on the per mu per time basis, which means that farm households began to adopt the resource quota system to allocate water. After only six months experiments, the farmland that paid water rates on the per mu basis made up 3% (Yahua Wang & Hu, 2002). The changes in the entitlement system provided the motive power for the changes in the degree of resource scarcity. The increase in the water scarcity helped push up the relative prices of water and stimulated the demand for new institutional arrangements for right divisions. It shows that the irrigation area in 1950 in the Yellow River Bend was 195046.6 hectares, but the figure shot up to 580626.67 hectares by 2000, three times as many as before 1949. During the same period, the water volume diverted from the Yellow River increased from 3.277 billion cubic meters to 5.158 billion cubic meters, less than twice as much as before 1949. The Yellow River Bend is an arid area, with the annual mean precipitation being less than 200 mm. It is therefore very costly for agriculture to depend on irrigation. The drop in the per mu water volume further pushed up the relative prices of water (Y. Wang & Tian, 2010). Resource scarcity, along with the large scaled construction of irrigation projects and the restoration of social order, has made it easy to understand why the irrigation management system changed so fast and why the resource quota entitlement system spread on such a big scale.

The operational cost of resource quota has been greatly reduced under the modern conditions, thus providing the necessary conditions for introducing the system. The cost of water measurement and oversight is the main reason for obstructing the introduction of the resource quota entitlement system. Only when the cost of water measurement and oversight is reduced significantly, is it possible for the changes in institutions in response to resource scarcity. Studies discover that, similar to the evolution of the ancient entitlement system, technical progress has played a crucial role in lowering the cost of water measurement and oversight. Compared with the modern water measurement technology, ancient water distance was extremely primitive and crude. Modern flow measurements include network of hydrological stations, water meter and velocity meter as well as such supporting systems as telecom facilities, power systems, automatic monitoring system and information processing system. They can accurately take the flow and volume of water on a large scope and under very complicated natural conditions.

Although modern technical progress has lowered the cost of the operation of the new institutions, the introduction of such technologies is also costly and should be regarded as part of the cost of institutional change, as the institutional change happens only when such cost can be paid. Statistics show that in 1999, there were 2912 water cross sections for measuring water in the Yellow River Bend area, all owned by the state, 407 steel frame water measuring bridges, 1427 steel reinforced concrete water measuring bridges, 78 sets of water meters and 631 sets of velocity meters. In addition, there are 5911 water measuring bridges, 284 sets of water meters and 593 pieces of instruments operated by the local people's management organizations. In the 1998–2000 alone, the water measuring supporting facilities cost 8,659,600 yuan. The size of investment in modern water measuring technology was unimaginable in the ancient times.

Under the conditions in which the water prices are raised, resource quota entitlement enjoys the advantage of economizing cost and that may explain the continued replacement of input quota entitlement with resource entitlement over the past 20 years and this trend is continuing, which is expressed in the spread of the water rate on the per mu and per time basis. Water metering on the per mu basis was initiated in the villagers group in Dongyang of the Xinsheng Village, Dongsheng Township in the Wulan front banner. The group has 52 households, 215 people and 2065 mu of irrigated land. With the rise in water rate toward the end of the 1980s, the irrigation areas began to measure flow volume to the public operated weirs of the delivery canal. In order that all the households share the water fees, they started in 1989 to introduce the one-time irrigation with each household as the unit that is, water is charged on the per mu basis and local farm households as counting the number of irrigation. The specific method of operation is like this: One water management person is responsible for the sequential order of irrigation among the villagers and records the areas irrigation each time. The wage for the person was 700 yuan, plus 300 yuan for counting the areas of land irrigated. After each round of irrigation ends, the actual areas of land irrigated would be verified and signed by the irrigators. When autumn ends, the total areas of land irrigation for the whole year would be added up together and the fees will be shared on the per mu basis. After nine years of operation, some households deemed it too trivial at the beginning of 1989 and each year, the irrigators have to give extra 300 yuan to the water manager. In the very year, the village began to share the water fees according to the land areas irrigated. The upshot is that the land that may or may not be irrigated was all irrigated, resulting in a 2.02 flow/day more water than in 1997 and the water fees increased by nearly 6000 yuan, 20 times the amount for the practice of counting the areas of land. In 1999, they drew on the lessons and resumed the practice of paying water rates on the per mu basis. This reduced the water volume used by 4.42 flow/ day as compared with 1998. Calculated by constant price, the water fees were 24,400 yuan less, with each household paying nearly 470 yuan less on average and each person paying 110 yuan less. The period of summer irrigation was also shortened by four days (YRHEO, 1995).

This example shows fully the function of institution in economizing cost. What the village gained was much more than the institutional cost (troubles and extra pay for water manager) and that is why the new institution was readily accepted by local villagers. A regression made the local people aware of the benefits brought about by the new institution. The fee collection on the per mu basis was established with more solid status. In addition, the Yellow River Bend Management Bureau became aware from the practice of the village that the fee collection on the per mu basis was a practice that could economize cost and water and very fair, thus having the value of being spread (YRHEO, 1995). After that, the bureau formulated preferential treatment for experiments within the irrigation area. A year's trial implementation achieved good results in many aspects that did not only save water but also cost. The experimental irrigation districts could at lease save water and reduce the burdens on farmers by somewhere between 10-26%. The irrigation areas were verified and water fees were share equitably. Excessive land reclamation was curtailed. Farm land capital construction went a step forward. After six months' demonstration, the irrigators became fully supportive to this method water measurement. All areas that began to implement this method spontaneously. Some areas even urged to accelerate the pace of spreading such method.

The difficult points for spreading this method water fee collection are highlighted in four aspects. One is about the cadres, as part of the townships, villages and cooperatives, were not active in accepting the method, because this method revealed the untrue areas of irrigated land, which had an adverse impact on other methods featuring sharing and levying of water fees according to land areas. Some places fumbled out the fair sharing of water fees in the past, making it difficult for the township and villages to settle the disputes. The second is that the fee collection on the per mu basis revealed the newly reclaimed land and 'black land' for which the water fees were not paid. For years, this part of land was not included in the areas calculated for water fees and they were mostly planted by village and cooperative cadres and 'black households'. When the irrigation areas were verified, the cadres and black households were not satisfied and some of them were very much influential, posing great resistance in spreading the new way of fee collection. The third is that there must be some input in fee calculation. There must be some volumetric facilities at the point of water diversion and people should be deployed to count the number of rounds of irrigation and they should be given a certain amount of pay. At the same time, if the work was done meticulously, the areas of land must be counted clearly and true to facts. Some people felt it troublesome and therefore were unwilling to introduce the method (Yahua Wang, 2013b).

The process of spreading the water fee collection on the per mu basis shows that this new institution became very popular, demonstrating its vitality. But just as our theories predict that institutional change has to pay not only static transaction cost (cost of the operation of the new institution) but also dynamic transaction cost (cost for the conversion between the new and old institutions). It is exactly the simultaneous existence of the two kinds of costs that have obstructed the spread of the new institution. Some places were unwilling to pay static cost, such as building facilities, paying wages to management personnel while some others were afraid of paying dynamic transaction cost, including losses resulting in other areas and the touching of the interests of interest groups. We can also see that the static transaction cost in institutional change is easy to resolve, such as by demonstrative projects to make people realize the gains of the new institution. But it would be more difficult to resolve the problem of dynamic cost, which is the main cause for the existence of efficient institution. Due to the unevenness of gains from region to region, that is, different places have different amount of transaction costs, the progress of institutional change is uneven. That will lead to inadequate supply of efficient institution only by inductive institutional change and therefore requires coercive institutional change. The Irrigation Area Management Bureau and local governments adopted such coercive means as publicity, mobilization, preferential policies and compensation to push this more efficient institution.

6.4.2 Explanation of Differences in Irrigation Water Entitlement Systems in the Upper, Middle and Lower Reaches of the Yellow River

There are significant disparities in the management systems between the irrigation areas in the upper and middle reaches and irrigation areas in the lower reaches, thus providing empirical materials for the horizontal comparison in the water rights structure in different areas. We have selected the Weishan Irrigation Areas using water diverted from the Yellow River in Shandong Province as a typical example for the lower reaches of the Yellow river. By studying the differences in the water entitlement systems between Weishan and the Yellow River Bend Irrigation Area, we shall further test the effectiveness of the theoretical tool developed in this book.

As it is pointed out in the preceding section, due to historical reasons, the water management institution in the irrigation area using water diverted from the Yellow River in the lower reaches fell behind that in the middle and middle reaches. Comparing the current irrigation management system of the two areas, we can discover that there are two major differences in the management system. The first is that the replacement of input quota entitlement system with resource quota entitlement system is far inferior in the lower reaches to that in the upper and middle reaches. The Weishan irrigation area supplied water free before 1979 and it did not collect fees according to the areas of land irrigated until 1980–1983. It was only after 1984 that it began to introduce the water supply by measurement and collect fees according to the water volume provided. At the beginning, the county (city, district) supplied water in a planned way by signing off tickets and later on, fee collection according to water volume spread to townships. Now 80% of the counties (cities and districts) and townships introduce the method and experiments are underway to introduce the method to the village level (Xu, 2002). The popularization of the resource quota entitlement in Weishan was still at the level of the late 1980s in the Yellow River Bend Area. The second difference is the degree of organization at the grassroots. In this regard, the irrigation area in the lower reaches was far lower than that in the middle and upper reaches. Besides, there are obvious differences in the path of institutional change in the grassroots water management. Water users associations get more attention in the irrigation area of the middle and upper reaches as such associations and 'association plus contracting' are the important models for the reform of water use at the grassroots level in Ningxia and Inner Mongolia (Hang, Zhongjing, You, & Calow, 2009; Xia & Pahl-Wostl, 2012). But in Weishan in the lower reaches, Water Users Associations have not received such attention as those in the upper and middle reaches and contracting, leasing and joint stock cooperatives are the main orientation of the property right reform of water used for irrigation at grassroots (Mao, 2002).

To understand the differences in the institutional arrangements with regard to irrigation, we need to know the different in the institutional environments of Weishan and Yellow River Bend area. The following three points merit attention. First, they are different in irrigation tradition. The Yellow River Bend Area started irrigation in the Han Dynasty and it has an irrigation history of more than 1000 years. Weishan, however, started irrigation only in 1958, barely 60 years up to the present. The second is that they are different in the way of irrigation. The Yellow River Bend Area is dry, with a mean annual precipitation of less than 200 mm and it has to divert water for irrigation. Weishan has a mean annual precipitation of more than 600 mm and irrigation is only supplemental. So their dependence on irrigation for agricultural development varies greatly. Third, they are different in economic development levels, with Weishan much better than in the Yellow River Bend Area. The income level of farming households is much higher in Weishan than in the Yellow River Bend Area and is the proportion of non-agricultural population, with the relative importance of agricultural production lower than in the Yellow River Bend Area. The three differences are universal, reflecting the differences in institutional environments in the irrigation areas between the upper and middle reaches and the lower reaches.

In terms of economics, the two areas have different environmental constraints and transaction cost systems in institutional change. First of all, the relative value of water for irrigation is much higher in the upper and middle reaches of the Yellow River than in the lower reaches and that has enable the upper and middle reaches to have a much stronger motivation to have more intensive management institution. Secondly, institutional change has to pay cost, which has been reduced due to its time-honored irrigation tradition, much lower than in the lower reaches that have no irrigation tradition at all. Therefore, the irrigation area in the upper and middle reaches has such cost advantage in introducing more intensive irrigation institutional arrangements. This may serve as an explanation why the replacement or input quota entitlement with resource quota entitlement is much higher than that in Weishan. The cost effectiveness also determines that the people in the Yellow River Bend Area have a higher self-organization level. In less-developed areas where irrigation is so important, village level has a much stronger collective action that is compatible with the local ways of production and life. The high self-organization capabilities have, conversely, helped reduce the cost of further institutional change. In the process of spreading the irrigation on per mu basis, water users associations have made it more convenient for the introduction of the new institution (Yahua Wang & Hu, 2002). But Weishan has to face fairly high cost in developing more intensive irrigation management institution and that determines the inadequate strength in collective action at the grassroots. Such low self-organization level will conversely raise the cost for introducing more intensive management institution. That is why Weishan is much slower than in the Yellow River Bend Area in measurement supply of water down to villages and even to farm households and in introducing other new institutions. The low self-organization level in Weishan has made the area more biased for the reform of property rights, because property rights reform has fairly low requirements for the grassroots collective action and the cost of opting for property rights reform is much lower than the cost of opting for water users association. This, in fact, is also an expression of the characteristics of 'path dependence' in institutional change (Krutilla & Krause, 2010).

6.4.3   Explanation of Changes in Other Entitlement Systems in the Yellow River Water Rights Structure

When the scope of water resource scarcity has been expanded, water rights allocation would be extended to cover great scope. There is the problem of options for entitlement system in the water allocation from the whole basin down to local governments and groups just as the water allocation in irrigation areas. The preceding section set forth the evolution over the past half century in the water rights structure for the use of surface water in the Yellow River basin and complete water allocation systems have been established at all levels. Figures 6.2 and 6.3 show perceptively that all the levels from basin down to local governments and from local governments down to groups have all opted for the resource quota entitlement system, that is, water is allocated among policymaking entities at all level. The following is an explanation of such option.

When water resource scarcity is not very serious, it is n0t necessary to impose resource quota. Water volume may be divided by restricting input quotas such as the number of water extraction projects. In granting entitlements to groups, it may not provide for the volume of water to be extracted. Some restrictions may be imposed on the exercise of water extraction rights. In fact, many river basins in southern China, including the Yangtze River, have adopted this entitlement system at the basin level. For instance, they restrict the water extraction capabilities through administrative examination and approval of water extraction projects. One of the advantages of input quota entitlement system is that it can result in the economical use of water and saving oversight costs. Under the condition in which water resource scarcity is not serious, input quota entitlement would always be the first choice. But with the aggravation of water scarcity, the drawback of the input quota entitlement system would reveal itself; due to lack of quantitative basis, it is unfavorable for settling the water use disputes among different localities and groups. That would motivate the transition from input quota to resource quota in granting entitlements, to realize direct allocation of water among localities and groups.

The Yellow River lacks water resources from the very beginning. Its average run-off is only 1/17 that of the Yangtze River. Water disputes arose in sections of the river from the very beginning of the founding of New China. In 1972, the lower reaches of the Yellow River suffered from intermittent interruption of flow and that happened every year in the 1990s. The severe water scarcity has made the Yellow River basin the first to introduce the water allocation scheme. So resource scarcity is the fundamental reason for the introduction of resource quota entitlement system in the whole river basin. Furthermore, modern hydrological technology has provided the material basis for water allocation. In fact, at the beginning of the twentieth century before the western modern hydrological technology was introduced into China, it was impossible to accurately measure the flow and water volume of the river; neither was it possible to allocate water among different areas. The water disputes in the river basin in ancient times was settled by limiting the water extraction capabilities of projects and fixed a rough proportion. After the founding of New China, the hydrological stations have developed steadily. The hydrological system has been brought to perfection since reform and opening up, and the cost for monitoring the run-off and water use has dropped steadily. The establishment of the water regulation information in recent years has made the costs of monitoring the amount of water extracted much lower. So we may consider that under the modern technical conditions, the water rights allocation according to entitlements among different areas and groups is an institutional option with the most economical transaction cast under resource scarcity. It can be predicted that with the aggravation of water shortage and information and technical progress, the entitlement system would change again to make resource quota allocation more accurate and more elaborate in the rules.

What merits close attention is that, the development of modern hydrological and information technology will further cut the costs of measurement and monitoring of the use of water resources, which will have a major impact on the institutional option and change. As far as the entitlement system is concerned, the cost of direct allocation of water will become lower and lower and resource quota will replace input quota at a faster speed to become the universal way of entitlement. The Water Law revised in 2002 has provided for the water allocation scheme based on basins, thus establishing the legal status of resource quota entitlement system. Besides, the hydrological and information technology development may cut the levels in water allocation. In the management system in the lower reaches of the Yellow River, the YRCC may directly allocate water to various weirs by the water regulating system to realize direct allocation from basin right down to group level. In other words, the progress in modern hydrological and information technology will objectively level off the water rights structure.

6.5   Empirical Analysis of Current Operation of Water Rights Structure

6.5.1   Current Water Resource Management and Water Rights Quality

Since the end of the 1990s, China has accelerated the reform of the water resource management system. The promulgation of the New Water law in 2002 marked the establishment of a complete set of water allocation system by administrative means. Figure 6.4 sums up the current water management system by using the hierarchy theory,with the water rights scheme at leach level divided into definition, enforcement and maintenance. The current water resource management system is the result of natural evolution since the founding of New China. In general, it is a set of administrative water allocation scheme, with the key institutions for defining the rights of policymaking entities at each level including macro regulation system at the basin level, the water allocation and total volume control scheme at the local government level, water extraction permit and quota management system at the group level and volumetric use of water and fee collection and additional price for

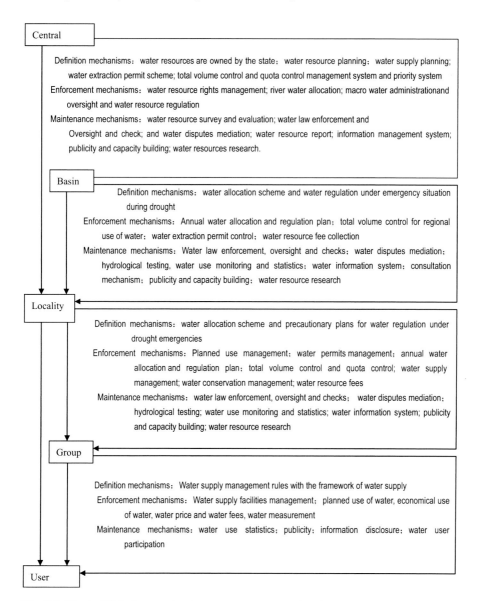

Fig. 6.4 Chart of China's current water resource management system

extra-water used system at the user level. It is not long to introduce The systems have not been introduced for a long time. Although they are embodied in all kinds of laws, corresponding enforcement and insurance mechanisms are lagging behind, so it is not guaranteed that regulations and management rules are well enforced (Boxer, 2001, 2002; Dukhovny & Ziganshina, 2011). The new Water Law provides that all units and individuals wishing to directly extract water from rivers, lakes or underground shall apply for water extraction permits according to the provisions of the state water extraction permit system and compensatory use of water resources and pay water resource fees and obtain water extraction rights. In 1988, China issued the water extraction permit system according to the provisions of the Water law and in 1993, the State Council issued detailed rules for implementing the water extraction permit system. In 1994, the YRCC, as authorized by the Ministry of Water Resources, began to issue water permits. In 1999, water permits were renewed, with the new permits being valid from January 1, 2000 to December 31, 2004. The YRCC also carries out annual checks of water permits. It can be regarded that a complete water extraction permit system has taken shape in the Yellow River Basin (Dukhovny & Ziganshina, 2011; Shen, 2004, 2014; Wouters et al., 2004).

Now let us look at the duration of water rights. The current water permit is valid for five years. During the period, if water extraction stops for one year or water extraction households have violated the law, their water permits will be revoked. This shows that the duration of water rights is relatively short in China. With regard to divisibility, the current water permit system has the constraints in these aspects, such as starting time and time limit, purpose of water extraction, water volume extracted, monthly water use within a year, insurance rate, water extraction method, water economizing measures, place of water recess and sewage water treatment measures. That means that water rights are brought under total control over time, space and purpose. Such strict external constraints have made water use hardly divisible. With regard to right transfer, the current water extraction permit system prohibits the transfer of water permits. So the current water extraction rights do not have transferability. Lastly, we look at the exclusiveness. The current management rules impose many responsibilities and obligations on water users, but there are no rules on the rights and interests of water users. Moreover, current law has empowered management organizations with the right to reduce and restrict water volume permitted to be used. Water volume reduction and restriction may happen in the following cases: (1) due to natural reasons, water resources are unable to satisfy the normal water supply in the localities; (2) geological disasters such as caused by excessive extraction of ground water or by land sink due to over-extraction of ground water; (3) total water extraction has increased but there are no other water resources available; (4) changes in products, output or production technology has resulted in the changes of the volume of water extraction; (5) other special situations that require reduction or restriction of water to be extracted. The above regulations have endowed management organizations with extremely big rights of intervention. Under the current institutional framework, the rights of water users are threatened by not only other users but also the nearly unlimited intervention by management organizations, thus making the exclusiveness of water rights very feeble. The analysis of water permit system is based on the detailed rules for the implementation of water permit system issued by the State Council in 1993. Although the rules are being revised by the State Council, the new management rules have not come off the table and the old rules are still valid. Besides, the 2002 new Water Law does not have the provisions about water extraction rights. The original management rules only have water permit system. The advancement of water rights is in itself a progress. It implicates greater respect for the water extraction rights. The water permit implementation rules now in the progress of revision are studying the problem of water rights transfer. So there might be relaxation in the rules of banning water right transfer. In a word, under the current institutional framework, the quality of water rights is very low, not only with rights holders but also the general water rights quality. According to the theories set forth in the previous chapters, the low quality of water rights is corresponding to the administrative allocation of water rights. So, it is certain that the current water rights quality is compatible with the administrative allocation of water rights in force.

Low water rights quality seems to have had inherent connection with the extremely strong state power. According to the statistics of 395 water permits holders (see Table 6.3), most of them are held by public organizations (nearly 90%). Within this, official agents of the state (local government departments and township government) made up 38.7% of the total and non-official agents of the state (villagers committees, water project management organizations) made up 41.8%. Others institutions, including water supply companies, water using institutions and state-owned enterprises are also controlled directly by the state. This shows that most of the water permit holders represent the will of the state or serve as an extension of the state power. There are virtually no groups independent of the state power. It can be imagined that overwhelming majority of the water projects and facilities are owned by the state, and they constitute the foundation for public organizations representing the will of the state to hold water rights while the rights held by group organizations formed for the interests of individuals (non-governmental organizations) are meager. They are far from being a force that can restrict the state power and lack the property right basis to exercise more rights on a larger scope. The characteristic of the distribution of water permit holders is a mirror of the current political and social structure and also a product left over from the period of planned economy. It will have a far-reaching impact on the institutional change (including water right transfer) in the water permit management.

6.5.2   Effectiveness of Current Water Rights Administrative Allocation System

As is known to all, a law in paper can never be automatically enforced. It requires a compatible enforcement and maintenance mechanisms and institutional arrangements, which go together to determine the performance of the law. China has gradually established an administrative water resource management system. However, the effectiveness of the system has to be investigated through empirical studies. This book uses a simple and clear method to study the effectiveness of the system, that is, observing the performance of the control targets of water volume actually used. If the data in time sequence of water actually used converge toward the control targets, it means that the administrative allocation is effective, with the convergence speed and degree reflecting the degree of effectiveness. Due to space limit and the availability of data, the author has selected the basin, local and group levels for empirical observation.

Observation of the basin level. Figure 6.5 records the natural surface run-off and the actual water consumption over the past half century. The figure shows that

Table 6.3 Classification of holders of water permits issued by the YRCC

Nature of water users	Local government departments	Township government	Villagers committee	Water project management organizations	Water supply companies	Public financed institutions	enterprises	Individuals
Number of water permits	91	62	82	83	15	15	27	20
percentage	23.0	15.7	20.8	21.0	3.8	3.8	6.8	5.1

Note: According to statistics of the 395 water permits issued in 2000 by the YRCC. Local government departments including river management bureaus; water project management organizations include irrigation management organizations; water using public financed institutions includes schools and hospitals
Sources: Sun Guangzheng, Qiao Xixian, and Sun Shousong: "Yellow River Resources Management", (see attachment Table 2: Registration of Water Permits Issued by the YRCC, pp. 472–539)

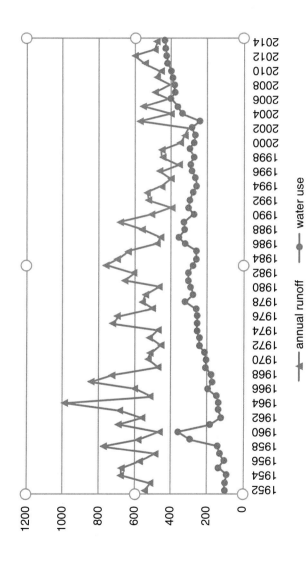

Fig. 6.5 Yellow River natural run-off and water consumption (1952–2014)

before 1989, the consumption of surface run-off in the Yellow River Basin rose steadily and its annual fluctuations changes in the opposite direction of the annual natural run-off fluctuations, which means that when there was a small flow of water (less rainfall), the water used increased; when there was a big flow of water (more rainfall), the water consumption was less. This is a natural water use pattern when water scarcity is not very serious. After 1989, great changes have taken place in the pattern of water consumption. In the whole of the 1990s, the natural flow of water tended to become less and less and the water consumption stopped increasing and was stabilized at about 30 billion cubic meters, indicating that the external artificial constraint in water consumption was intensified, alienating the water process from its natural pattern. Further observation of the two curves in the 1990s shows that, except some particular dry years (1991 and 1997), the annual fluctuation of water use is in the positive direction with the annual fluctuation of natural flow of water, which means more water consumption with more water is available and less water consumption with less water available, reflecting the water use characteristics when water scarcity is serious. Since 2003, the Yellow River basin has entered a period of abundant water, with the water consumption increasing steadily.

The above change in the water use pattern reflects the functions of the water resource allocation system at the basin level. In 1987, the Yellow River basin came out with a water allocation scheme, which was enforced by an enforcement mechanism in the whole of the 1990s, especially after 1997 when the scheme was strictly enforced. Although there is no way of accurately evaluating the effectiveness of the water allocation system in the Yellow River basin, the fact that the run-off consumption was effectively controlled at a stable level has proved the effectiveness of the water allocation system at the basin level.

In terms of the local level, Table 6.5 shows the actual water consumption in the 27 years after the eight riparian provinces and regions enforced their water allocation schemes and the comparison with the 1987 allocation norms. The table shows that six provinces (Gansu, Shanxi, Shaanxi, Henan, Qinghai and Ningxia) did not use up the water quotas and overused their quotas in some years. The two others (Inner Mongolia and Shandong) overused their quotas in most of the years. In general, Ningxia, Inner Mongolia and Shandong hit the allocated norms and currently they do not have additional quotas. The overused water volume of the three places is drown in Fig. 6.6 and discovered that although the proportion of overuse in Shandong and Inner Mongolia in the whole of the 1990s was bigger than zero, its changing trend all converges toward the horizontal axis. By 2000, it basically met the quota and in some years after that it was below the quota allocated. What is worth special mention is that since 1999, Shandong overused its quota in part years while Inner Mongolia used water below quota in 2003, being the first time in history. This process shows that the water allocation system among different provinces and regions has not only played its roles but also been intensified with the passing of the time.

Lastly we shall examine the effectiveness of the water allocation mechanism at the group level. Authorized by the Ministry of Water Resources, the YRCC is responsible for the issue water permits and monitoring of its implementation in the

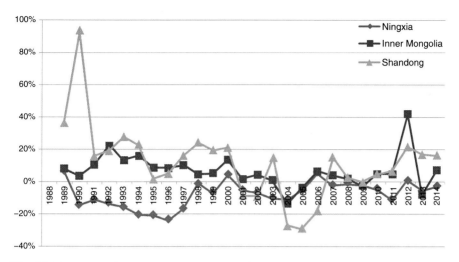

Fig. 6.6 Percentage of water overused by Yellow River riparian provinces (1988–2014). Notes: Calculated based on the data in Table 6.4

whole of the Yellow River basin (Y. Wang and Tian, 2010). Local governments are responsible for managing water permits beyond the scope covered by the basin management organization. Due to availability of data, this study is confined to water permits issued by the basin management organization. The YRCC completed the first registration of water permits in 1996 and then strengthened monitoring and management. In 2000, it carried out the renewal of the water permits. The author has collected data about 200 water users issued by YRCC and calculated their extra-quota water extraction ρ in the four years of 1997, 1998, 1999 and 2001. ρ is defined as follows:

$$\rho_i^j \equiv \omega_i^j \Big/ _{\varpi_i} - 1 \quad i = 1, \ 2 \ \ldots \ 200; \quad j = 97, \ 98, \ 99, \ 01$$

ω_i^j = actual water extracted in j years by user i; ϖ_i is water quota of user i approved to use 2000 water permits approved. When $\rho > 0$, shows quota overused; when $\rho \leq 0$, shows water used within quotas. Theoretically, the value of $\rho \in [-1, \infty)$. $\rho_i^{97}, \rho_i^{98}, \rho_i^{99}, \rho_i^{01}$ is reflected in Fig. 6.9, in which $i = 1, \ 2 \ldots 200$ is by and large arranged in the order from upper reaches down to the lower reaches. Water users whose $i > 51$ are in the lower reaches (Henan and Shandong). Figure 6.7 shows: (1) the data dots are very unevenly distributed on the horizontal axis, reflecting the unevenness in the allocation of water quota among water users; (2) most of the data dots each year are distributed below the horizontal axis, indicating that most of the water users have not used up their quotas; (3) the scale maps of the four years vary greatly, indicating the big changes in the annual water volume extracted by users; (4) almost all the data dots exceeding 0.5 are distributed in the lower reaches and those exceeding 0 are distributed in the lower reaches, indicating that users that have overused their quotas are in the lower reaches.

Table 6.4 Actual surface water consumption in provinces and regions in the Yellow River Basin (1988–2014) Unit:Billion m^3

	Qinghai	Gansu	Ningxia	Inner Mongolia	Shanxi	Shaanxi	Henan	Shandong
Allocated quota	1.41	3.04	4.00	5.86	4.31	3.80	5.54	7.00
1988	1.34	2.57	4.25	6.33	2.21	1.21	5.09	9.55
1989	0.98	2.33	3.41	6.06	1.96	1.44	3.72	13.48
1990	1.00	2.36	3.54	6.46	1.85	1.23	3.30	8.09
1991	1.59	2.39	3.46	7.16	1.97	1.25	4.00	8.32
1992	1.58	2.45	3.37	6.62	2.10	1.32	3.38	8.93
1993	0.99	2.11	3.17	6.79	1.73	1.03	3.55	8.61
1994	1.06	1.76	3.16	6.35	2.28	0.92	2.88	7.11
1995	1.07	1.84	3.04	6.35	2.20	0.90	3.12	7.33
1996	1.35	3.00	3.32	6.44	2.21	0.99	3.77	8.12
1997	1.2	2.58	3.95	6.12	1.21	2.50	3.67	8.70
1998	1.16	2.35	3.71	6.15	1.97	1.05	2.95	8.36
1999	1.21	2.58	4.15	6.65	2.09	0.96	3.46	8.45
2000	1.32	2.74	3.78	5.95	2.18	0.99	3.15	6.39
2001	1.126	2.692	3.7	6.103	1.046	2.178	2.942	6.341
2002	1.169	2.612	3.574	5.918	1.043	2.111	3.601	8.032
2003	1.089	2.917	3.559	5.046	0.960	1.873	2.825	5.057
2004	1.062	2.933	3.767	5.639	2.091	1.007	2.607	4.957
2005	1.077	2.921	4.208	6.22	2.36	1.181	2.932	5.73
2006	1.357	3.005	3.902	6.094	2.684	1.29	3.777	8.046
2007	1.333	3.044	3.944	5.97	2.497	1.358	3.364	7.159
2008	1.211	3.011	3.896	5.708	2.68	1.447	3.943	6.966
2009	1.104	2.991	3.798	6.134	2.558	1.508	4.336	7.336
2010	1.051	3.032	3.547	6.129	2.442	1.817	4.41	7.449
2011	1.215	3.721	4.027	8.314	4.537	3.903	6.53	8.496

2012	0.902	3.188	3.755	5.394	2.772	2.066	5.386	8.162
2013	0.944	3.089	3.885	6.275	2.95	2.208	5.323	8.133
2014	0.932	2.993	3.88	6.2	2.948	2.323	4.677	9.246

Source: Compiled based on the "Yellow River Water Resource Bulletin" for years (1988–2014), YRCC

Note: Data in the first row are the quotas for all provinces and regions by the State Council 1987 water allocation scheme

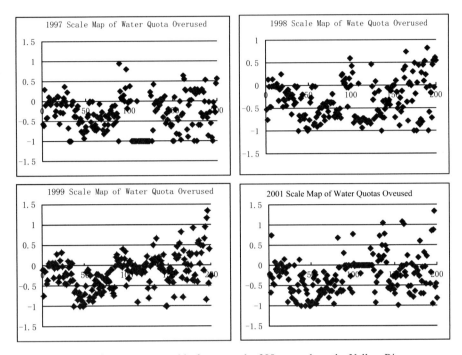

Fig. 6.7 Changes in water extracted in four years by 200 users along the Yellow River

In the following we shall make more detailed analysis of the sample data. According to the survey conducted by the author, before the water permits were renewed in 2000, the oversight and management of water permits were very weak and the external artificial constraints on water users were small, so the use of water in 1997, 1998 and 1999 was close to natural course of development. However, after water permits were renewed in 2000, with the unified regulation strengthened, all places have intensified the management of planned use of water at all weirs and so water extraction in 2001 should be different. We use the data for 1997, 1998 and 1999 to observe the annual fluctuations in water extraction under natural conditions. The fluctuation co-efficient σ is defined as follows:

$$\sigma_i = \tfrac{1}{3} \times \left(\sum_j \left| \rho_i^j - \tfrac{1}{3} \times \sum_j \rho_i^j \right| \right), \quad i = 1, \ 2 \ldots 200; \quad j = 97, \ 98, \ 99$$

Based on results calculated according to σ, it has arrived at the following conclusions: (1) there were big fluctuations among all water users and the average value of σ is 0.185, which means that the water volume changed in the three years for every water extractor on average at 18.5% of the quotas; (2) the fluctuations of 37% of the water extracting households exceeded 20% of the quotas; (3) the fluctuations in the lower reaches were much bigger than in the upper and middle reaches. The sources of fluctuation were very complicated, with influencing factors including weather conditions, economic structure, method of water extraction and

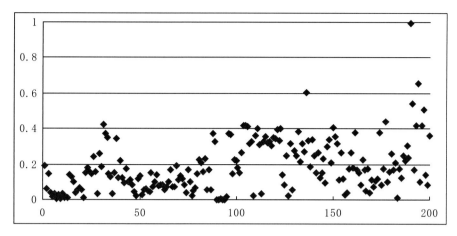

Fig. 6.8 Fluctuations of water extraction by 200 households along the Yellow River

Table 6.5 Water extraction before and after water permits re-verified in 2000

Year	Total water volume saved (billion m^3)	Households that used to economize use of water have overused water	Households that used to overuse water have economize use of water	% of households that used to overuse water have economized use of water (%)
1997–1999	−1.227	34	19	39.6
1999–2001	3.808	28	41	60.3

Source: Calculated by the original data provided by the YRCC, with 200 water permits samples

water delivery facilities. Figure 6.8 reflects perceptually the fluctuations of water extraction of all water extractors in 1997–1999. It should be mentioned in passing that from Fig. 6.5 we can feel perceptually the violent fluctuations of natural flow of water. Such hydrological uncertainties have been familiar to all. But the fluctuations of water use reflected in Fig. 6.8 often elude the people's attention. Such uncertainty in the use of water would no doubt affect the option for water management institutions.

In order to examine the effect brought about by water permit management in 2000, the author has made comparative analysis of the changes of ratio ρ of extraquota water extraction in 1997–1999 and in 1999–2001 and found that there were diametrically different changing trends during the two periods as shown in Table 6.5. In the 1997–1999 period, the number of water extraction households whose ρ changed from negative to positive was 34; but in the 1999–2001 period, the figure dropped to 28. If we say that the change is too small to prove anything, the change in the number of water extractors whose ρ changed from negative to positive was very obvious. There were 19 in the first period and 41 in the latter period and among all the water extractors whose $\rho > 0$ at the initial year, the percentage of those whose ρ value was negative rose from 39.6% to 60.3%. Although the preceding section pointed out the fluctuating characteristics of water use, the changes in these data have obviously gone beyond the scope of natural fluctuation, attributable only to institutional change. The changes in the total water economized as shown in Table 6.5 are a further illustration of the apparent differences between the two-time period, with the water volume extracted in the former period increasing by 1.227 billion cubic meters and that of the latter period economized by 3.808 billion cubic meters. Although the weather and technical progress factors cannot be ruled out to account for the change, the strengthened management of water permits is no doubt the most important reason in inducing such changes.

The above has made empirical testing of the water allocation mechanisms at the basin, local and group levels, indicating the effective side of the water allocation by administrative means. Though our testing is not very stringent, it is enough to prove the administrative means, such as river water allocation, total water volume control of water use by various regions and the water permit system, which have indeed played its rules in the development and utilization of water resources. Meanwhile, this test shows that the current water allocation by administrative means is not so satisfactory. Although the water right allocation mechanisms have played its roles, but the actual use of water is entirely matched with allocation schemes, as there were still some policymaking entities failed to control their water use within the quotas allocated. The actual use of water at all levels deviated a great deal from the water use quotas. That has made people to cast doubt over the rationale of the initial allocation of water rights. But can more rational initial allocation of water rights solve the above problems? In real world, it is impossible to have a perfect initial allocation. Even if the initial allocation scheme is perfect, it has to be of necessity adjusted with the passing of the time and the changes in the water use process. The key to the problems lies not in the initial allocation but in an effective water rights reallocation mechanism.

Is the current water rights reallocation mechanism effective? The answer is positive. First of all, the empirical study of the Yellow River basin shows that water rights adjustments by administrative means may operate effectively with regard to the water rights for irrigation within groups. For instance, in the 20 years since reform, there were four adjustments in 1981, 1986, 1991 and 1998 between the general trunk canal and sub-trunk canals in the Inner Mongolia irrigation area according to the irrigation area, project conditions and crop plant structure. The adjustments have effectively maintained the balance of water use in the irrigation area and the growth of irrigation efficiency (Hang et al., 2009; Yahua Wang, 2013b). I further discover that the water allocation adjustments were more frequent among branch canals down below the sub-trunk canals and it was adjusted every year in the water volume for irrigation among farm households. Secondly, the current management system has the ability to make water rights allocation more rational by adjustments at the level above groups. For instance, the current water permits are renewed at a five-year interval and annual checks are carried out. All these are ways for improving the water rights allocation. The actual amount of water extracted by 200 water extracting units analyzed above is quite different from the amount of quotas allowed and it is entirely possible to make it more rational through further adjustments by administrative means. After all, it a very short time since the water permit system was introduced and there are still rooms for improvements. Although it is more difficult to operate in the adjustment of the water allocation schemes among different localities, it may be, in principle, revised under the sponsorship of the State Council. In a word, there are still problems with the current water allocation. Though all can be resolved theoretically under the current institutional framework, there are difficulties in practice. However, they can be improved if the laws and regulations are to be effectively enforced and implemented.

6.5.3   Water Rights Institutional Change (1987–2002) in the Yellow River Basin

It can provide a special revelation to review the institutional change of water rights in the Yellow River basin in the year of 1987–2002, when it is in a hydrological cycle of sustainable drought duration. Since the Chinese reform program, the water rights structure for the use of surface water in the entire basin has experienced drastic changes. In the development and utilization of water resources, people always tend to capture more rights and such capture acts are only constrained by water resource conditions and the costs of water extraction technology, without any artificial institutional constraints. It can be imagined that the open access of water resources would lead to drastic growth in water consumption. That is why there were intermittent dry-up of the river course in some sections of the lower reaches of the Yellow River in 1972. The dry-up of the Yellow River Course in the lower reaches forced the central government to set about studying the water allocation scheme. The implementation of the water allocation scheme in 1987 marked the introduction of an official water allocation rule, ending the history of disorderly use of water in the riparian provinces. It also marked the establishment of water allocation mechanism at levels above groups. That has never been seen in history and that is the first water allocation scheme since the founding of New China. As the water resources of most of the rivers and lakes in China have not been allocated, the practice in the Yellow River since 1987 has provided valuable experience.

From the water allocation, the author has discovered that the use of the framework provided by the water rights hierarchy conceptual model is not enough to explain the institutional change in the Yellow River basin during 1987 to 2002, as the water allocation scheme marked the establishment of water rights allocation mechanism. The dry-up situation in the course of the Yellow River in the lower reaches has further been exacerbated. That requires further exploration into such classification of entitlement system, initial allocation mechanism and re-allocation mechanism, which introduce more detailed classification of water rights regime. Table 3.4 provides a more general analytical framework, indicating that the water rights regime is a set of rules with correlationship and mutual complementarity, including three categories of institutions and nine categories of mechanisms, that is, the allocation system, enforcement system and maintenance system, initial allocation system, re-allocation system, ad hoc adjustment system, monitoring mechanism, incentive mechanism, punitive mechanism, information mechanism, interest integration mechanism and insurance mechanism. The three kinds of institutions in the water rights hierarchy conceptual model are only the allocation systems, without covering the enforcement system and maintenance system. We use the framework provided in Table 3.4 to examine the water allocation practice in the Yellow River basin.

The water allocation at the basin level during 1987 to 2002 may be divided into two periods: the first period is from 1987 to1997. In 1987, the State Council issued the schemes for the allocation of water available in the Yellow River to define the initial rights to use water for riparian provinces and regions. In 1994, the water permit system went into more details in the initial allocation of water rights. The second period covers 1998–2002. In 1998, the Ministry of Water Resources and the State Planning Commission jointly issued the annual allocation of water available in the Yellow River and the schemes for flow adjustments during dry seasons. In 1999, the YRCC started basin-wide unified regulation of water resources according to the rules on the management of flow regulation of the Yellow River.

The institutional change of the Yellow River water allocation can be regarded as a process of feeling the stones in wading across a river during the time from 1987 to 2002. It can be divided as two periods. The first period was an introductory period from 1987 to 1997, which the institutional designs were full of flaws. First, the allocation mechanism was not complete and the initial rights definition was not rational and water permits management was not strict. The second period from 1998 to 2002 is that volumetric allocation which lacked the support of enforcement system, especially the integration mechanism. The result was that the water volume allocation was not extensive approved. The third is that there was no enforcement system for volumetric allocation, the weakest point in the institutions. It is hard for water rights allocation to realize automatically and there must be well matched with corresponding monitoring, incentive and punitive mechanisms. In the ten years following 1987, the water allocation schemed was difficult to implement. Some provinces went even to violate the rules to over-extract water. This reflects the institutional failure, which lies in the defects in the design of the water rights institutions (A. Hu & Wang, 2000).

When an institution could hardly maintain itself, it would feel the tremendous pressure for institutional change. The series of reform measures after 1997 was a response to the crisis of Yellow River course dry-up. The second period saw a lot of new moves in water allocation within a very short period of time. This was a period of improvement in the water allocation system. First, the allocation mechanism began to improve, with a view to making the initial allocation more rations during rainy and dry seasons and the water permit system began to take shape, with the right division and being more detailed and clearer. Second, an enforcement system began to emerge as the importance of unified regulation has been accepted extensively. The YRCC set up regulation and enforcement department to use

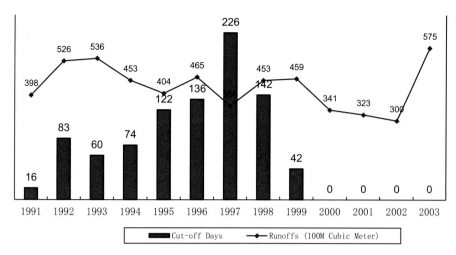

Fig. 6.9 The natural runoff and the number of cut-off days in the Yellow River. Source: sorted out and compiled based on the Yellow River Resource Bulletin by the YRCC

administrative means to monitor the regulation. Third, the maintenance mechanism was strengthened. The basin organization strengthened monitoring and research. After 2000, no river course dry-up appeared in the Yellow River, which may be regarded as an achievement in improving the water allocation system.

Institution is a group of complete rules, with each rule associated and complementary to each other. The effectiveness of institutions is determined by the changes in the whole institutional system. The institutional arrangements in the Yellow River water allocation over the past 15 years show that the knowledge about the completeness of institution has been enhanced. From the introduction of single institution to the improvement of the whole institutional system and from the flawed institutions to improvement, the institutional began to develop from its role as total soft constraints to a role of hard constraints. The 13 consecutive dry years since 1990 have sharpened the contradiction between water supply and demand as seen from the water use angle, but the bad things have turned into good ones as seen from the angle of institutional arrangements. The successive dry spell has helped deepen the understanding of the importance of institutional arrangements and thus accelerated the pace of the change in the water rights institutions in the Yellow River basin. Figure 6.9 reflects the big gaps in the effectiveness of water rights regime before and after 1998. The same years when water volume sharply decrease, the Yellow River flow was interrupted for 226 days in 1997. From 2000 to 2002, there was not flow interruption at all, even though the extraordinarily dry weather lasted for three successive years.

The practice of water allocation in the Yellow River basin has enriched our understanding of the water rights regime. The successive flow interruption in the Yellow River before the 1990s shows that it is far from being enough to formulate water allocation schemes. With the compatible enforcement and maintenance mechanism, the allocation scheme could only serve as soft constraint or even remained on paper only. Water rights regime is a system with associated institutions. The three kinds of institutions in the water rights hierarchy conceptual model are only part of the whole regime. The performance of the water rights regime depends on not only the introduction of individual rules but, to a large extent, on the compatibility and adaptability of the individual rules. Due to the complex nature of the water rights regime, all individual institutions usually change in a balanced manner along the path for induced institutional change. If the rules are well compatible, the performance of the water rights regime is usually high although the changes in water rights institution may be slow. Along the path of forced institutional change, some institutions may be introduced quickly; however, the supporting institutions often lag behind and are even ignored, and the incompatibility of the institutions may stand out. Although the change may be superficially fast, the actual enforcement results are not ideal.

The process of the water rights institutional change in the Yellow River basin has revealed a unique crisis-driven reform path. There have been two major changes in the water rights structure since reform, all driven by the crisis of flow interruption of the Yellow River. They are forced institutional changes under the sponsorship of the central government. The crisis or problems with the water resources in the Yellow River is, in fact, the lagging effect of water rights institutional change, which may be regarded as a consequence of inadequate institution supply. In this hierarchy water governance structure, administrative means is a way that can minimize the cost of maintaining the order of the water rights regime. Even so, the sound operation of the water rights regime needs administrative control and a high management cost on the part of the state. In the absence of government, that is, the state fails to pay management costs required in maintaining the structural stability, and it would lead to instability of the governance structure and the degree of hierarchical structure would drop. The bigger the shortage of management costs is, the more outstanding is in the government absence. The water rights structure would be more manipulated by natural forces, which will cause water disputes between upper and lower reaches, over-extraction of water for maintaining ecological balance, waste of water and low efficiency in the use of water. These consequences would add up to cause a so-called 'water crisis'.

Chapter 7
Water Rights and Water Market: Case Study in Contemporary China

This chapter studies the developments of water market in contemporary China by the case study method under the water rights hierarchy framework. Six cases have been selected, all taking place over the past 15 years, which reflect the latest changes in the water rights structure. This chapter aims to reveal theoretical implications of these cases. First of all, the author presents an analytical framework for the case study and divides the six cases into five categories according to the water rights hierarchy conceptual model. The first category covers water rights trading at the user level as depicted by the case in the Hongshui Irrigation Area of Minle County in Northwest China's Zhangye Prefecture, with the irrigation water rights trading in the Guangzhong Area during the Qing Dynasty as a contrast for comparative study. The second category deals with water rights trading at the group level as shown in the case an industrial firm buying agricultural water rights in the Yellow River basin. The third category is about water rights trading at the local government level, with the Dongyang-Yiwu long-term water rights transfer and a short-term water rights transfer, in the upper reaches of the Zhanghe River. The fourth category covers what is known as water bank that is regulated and controlled by superordinates, which is still in the process of planning in the Yellow River basin. The fifth category is the initial water rights allocation by the market within the plan of South-to-North water diversion project. The first four cases are the emphasis of the study in this chapter, as they are all rights trading among decision-making entities at the same level. The chapter winds up with a summary of the cases.

7.1 Analytical Framework of Case Study

The preceding two chapters examine in detail the changes in the surface water rights structure of the Yellow River, which is representative of the changes in the surface water rights structure in China. At present the run-off is not allocated in

© Springer Nature Singapore Pte Ltd. 2018
Y. Wang, *Assessing Water Rights in China*, Water Resources Development and Management, DOI 10.1007/978-981-10-5083-1_7

most of the major rivers. The Yellow River is one of them among the major rivers that has its basin-wide water allocation scheme. Despite all these, the 2002 new water law has fixed a series of water management systems, to a large extent, based on the experience of the water resource management of the Yellow River. The 'Yellow River Model' has become universal through codification. Although the water rights structure of the Yellow River has a certain degree of being advanced in nature, it represents the current level of China. In practice, most of the rivers need institutional changes and need to establish water allocation mechanism at the levels above the group. Figure 7.1 is a chart of the rights structure of surface water during the period of planned economy. It shows that the use of water was in a state of open access at the basin and regional levels during the planned economy period, when water rights allocation mechanism was established only within groups. The initial allocation and reallocation of water rights were all done by administrative means. This fitted the conditions of the planned economy. As the utilization of water by groups lacked external constraints, such water rights structure would inevitably lead to extensive expansion in the use of water. Due to lack of incentives for economizing on water use, the water use efficiency by users was low, resulting in exacerbation of water scarcity, which occurred from the group level up to the regional and even the basin level.

The market economic transition and the economic and social development have stimulated the establishment of water rights allocation mechanism at higher levels. After the Water Law was promulgated in 1988, the process has been accelerated and such mechanism has been brought to perfection gradually. Figure 7.2 shows the current water rights structure. It shows that, after three decades of development, allocation mechanism has been established at all levels. For instance, all the

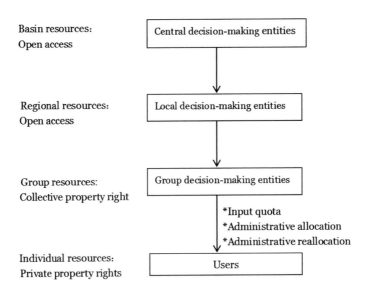

Fig. 7.1 Rights structure of surface water in the period of planned economy

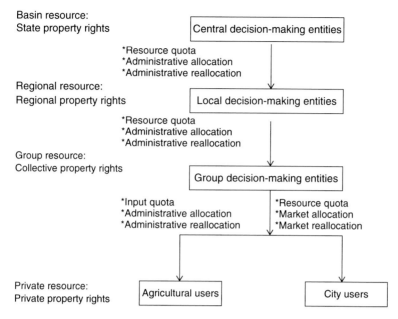

Basin resource:
State property rights

Central decision-making entities

*Resource quota
*Administrative allocation
*Administrative reallocation

Regional resource:
Regional property rights

Local decision-making entities

*Resource quota
*Administrative allocation
*Administrative reallocation

Group resource:
Collective property rights

Group decision-making entities

*Input quota
*Administrative allocation
*Administrative reallocation

*Resource quota
*Market allocation
*Market reallocation

Private resource:
Private property rights

Agricultural users

City users

Fig. 7.2 Current rights structure of surface water in China

riparian provinces in the Yellow River basin have defined their water rights according to water allocation agreements and allocated water withdrawal rights among groups based on the water withdrawal permit system. The current water rights structure is by and large a complete administrative control system. Water rights are allocated at both the initial stage and the adjustment stage by administrative means from central down to the group. The option for the current water rights allocation system falls in with the understanding of China's hierarchy water governance structure presented in this book.

The study has also taken notice of the fast process of industrialization and urbanization and the transition in the economic system. In cities and towns, there are already daily expanding water supply markets. This indicates that the current water rights structure of China has already deviated from the pure hierarchy structure. Market mechanism has been introduced into part of the administrative allocation system. This is a brand new development in the current water rights structure (see Fig. 7.2). This book has made a brief explanation that in an increasingly liberal environment, the costs of market supply of water in cities and towns are rapidly dropping and the degree of market allocation is rapidly increasing. In contrast, as the measurement and monitoring costs of irrigation water allocation are much higher than water supply in cities and towns, the sole dependence on the market allocation of agricultural water is costly and therefore the traditional administrative allocation remains.

As irrigation remains the main water user in China (63.1% in 2015). The current rights structure of the surface water use remains an administrative hierarchy system, with the administrative allocation and reallocation of water occupying a dominant position. During the transition to the market economy, the costs of administrative allocation rises and so are the costs of the operation of the hierarchy. Thus, it is possible to generate motivation for the introduction of market elements driven by the desire of lowering the costs of the operation of the governance structure. Over the past three decades, such motivation has driven the market water supply in cities and towns to a higher level. This process has been continuing over the past 15 years. Market mechanism has been expanding into the area of irrigation water use. Market allocation has also applied at higher levels. All these point to the direction of introducing market mechanism into the water rights structure. What is the nature of the cases of water market that happened successively? Why has it happened in an administrative allocation system? What is the accurate implication of the change in the water rights structure in China? The following is an analysis of the typical cases in a bid to answer these questions.

This chapter uses six typical examples of contemporary water market and five of them happened in the past 15 years. As water market is new in contemporary China and the systematic research on them is still little. In studying the cases, the author carried out field investigations into four of the six cases and collected a large amount of first-hand information. I try to analyze these cases under the theoretical framework provided by this book and unveil the underlying motivation and policy implications behind them. Six cases are divided into five categories as shown in Fig. 7.3, which are all at different levels of the water rights

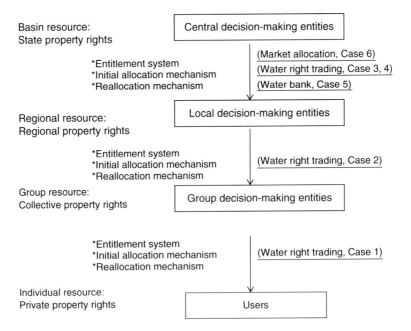

Fig. 7.3 Classification of contemporary water market cases

structure, with particular features for each case. The cases will be analyzed from a down-up order.

7.2 Water Rights Trading at User Level

The study of the irrigation management system of the Yellow River in the last chapter shows that in most cases, farmers pay water fees according to the area of crops irrigated and management organizations allocate water in a planned way among channels and canals above the branch weirs. In such a water allocation pattern, the quality of water rights held by farm households is not high and it is not easy for them to trade their rights. The water volumes used by them are mainly adjusted by administrative means between weirs. The water reallocation capabilities down the weirs are usually very weak. The irrigation water allocation is different from urban water supply because irrigation water and urban water have different characteristics: (1) Different in the physical nature of the urban water supply pipelines and the irrigation canal networks;[1] (2) Urban water makes it convenient for users to measure the water volume used while it is difficult for users of irrigation water to do so. (3) Urban water users have different fee bearing capacities from farm household users of irrigation water. If the three differences can be eliminated in irrigation, it can be imagined that irrigation water can be allocated by the market all the same just like the urban water supply. In the water supply market, there is trading between users and suppliers but not among users. So the water rights trading at the user level is confined to irrigation water. Trading in irrigation water rights began to happen in the Ming-Qing periods. Chapter 5 of this book makes a detailed analysis and has found that it was the result of the improvement in the irrigation water rights quality in the long historical period. But in contemporary China, the quality of irrigation water rights held by farm households is generally very low, making it very difficult for water rights trading to happen. An example is the experiment of building water-saving society in Zhangye of Gansu Province located in the northwest China.

Case 1: Water Right Trading in the Irrigation Area of the Hongshui River in Minle County of Zhangye in the Heihe River Basin
Zhangye in Gansu Province is the first area in China to experiment on the building of a water-saving society. The experiment started in August 2001 in the Liyuan

---

[1]This difference leads to the different characteristics of urban water and irrigation water. In cities, many users can obtain water from the same water supply pipelines while irrigation water users have to get water according to the sequential order. This is very important, because it determines that urban water supply makes it easy for introducing market mechanism but planned way is stressed in the use of irrigation water and it needs large scaled collective cooperation. Historically, such special features of the irrigation system had a great bearing on the development of the oriental civilization.

River Irrigation Area of Linze and the Hongshui River Irrigation Area of Minle. The Hongshui River Irrigation Area uses natural flow from reservoirs for irrigation, covering an area of 322,000 mu and a population of 97,000, with the cultivated land averaging 3.6 mu per person and the annual volume of water used is 110 million m^3. In October 2001, the Area is issued water withdrawal permits (valid for five years) and initially allocated water rights. After that, water rights trading took place inside the irrigation area.

Zhu Hong is a peasant of the Pengzhuang Village in the irrigation area. He has a family of four, tilling 14 mu of land. Before 2002, he mainly planted grain and used nearly 160 m^3 of water per mu in each round of irrigation and the annual water fee was 750 yuan. Before 2002, he said, the irrigation water was managed by village cadres and the management was rather lax. It often happened that households whose land was in the upper reaches of the river used more water than those in the lower reaches, and those with bigger influence used more water than those with less influence. Just like the egalitarian practice known in China as "everyone else eating from the same pot", it was not definite as how much water can be used for a mu of land. This resulted in the overflow of the channels and flooded irrigation, causing great waste, and everyone showed indifference about it. It was a universal phenomenon of 'free riders' in the collection of water fees.

After the water withdrawal permits were issued and the peasant water users association was set up, great changes have taken place. Zhu Hong got the allocation of nearly 4000 m^3 for his 14-mu land, much less than the volume actually used before. He adjusted the plant structure so that all the land is irrigated, thus achieving significant result in saving water. In the first round of irrigation this year, his water rights were 800 m^3 and he used only 600 m^3 to finish the irrigation of his 7-mu wheat and beer barley, saving 200 m^3. He sold the water saved for 0.2 yuan per m^3 to his fellow villager Sun Kairong as against the 0.1 yuan as fixed by the village. In the second round of irrigation, he had 100 m^3 short of the requirements for his 14-mu land. He then bought 100 m^3 at a price of 0.1 yuan per m^3 from his brother.

The account he did shows that he bought water rights for 80 yuan in the first round of irrigation and sold the water rights saved for 20 yuan. So he actually spent 40 yuan on the irrigation of his land. In the second round of irrigation, he spent an additional 10 yuan buying water rights. After the end of the second round, he saved 10 yuan in water fees. As seen from the whole year, he said, he may have used up all the water rights. However, through water rights trading, he has ensured the water requirements in each round of irrigation, unlike in the past when there were 304 mu of land that could not be irrigated in every round of irrigation. After the allocation of water rights, he has got clear about the amount due to him and the annual water fee payment has also become transparent—400 yuan for 4000 m^3 of water, 350 yuan less than before the new system was introduced.

Chao Huaipu, president of the water user association of Pengzhuang Village, said that the total amount of water rights transaction has reached nearly 10,000 m^3. The water rights trading has not only ensured water supply to all the land in the *village but also helped realize the dynamic balance of the total volume. In addition, the total amount of water saved came to 100,000 m³, as compared with the past.*

It was learned that water rights transaction in the Hongshui River Irrigation Area mainly occurred in the following circumstances: (1) when water is saved after structural adjustments; (2) water loss is reduced due to scale planting by connecting up the land; (3) when farm families are unable to grow crops due to lack of labor; (4) water is saved after water economizing engineering measures are adopted; (5) water rights trading under the regulation and control by the government. The government encourages structural adjustment and scale planting or rewards advanced workers in irrigation project construction.

Local farmers deem the water rights regime and water rights trading to have the following advantages: (1) Enhance the awareness of villagers in economizing on the use of water. As every family knows how much water can be used, they have become meticulous in counting the numbers with regard to water use; (2) promote structural adjustments. By promoting structural adjustments, local peasants can achieve both economical use of water and increased production—hitting two birds with one stone; (3) change the ways of using water. The egalitarian way of using water has become a history; (4) it has become more transparent in the use of water and free riders in fee collection have disappeared and water fee problem is no longer a main contradiction in the rural areas and the relations between cadres and common people have also improved (source: interviews and notes during fieldwork).

According to the description in Case 1, water rights trading is a product of the pilot project, but it has underlying motivation. In the author's view, the original driving force behind the water rights trading is the scarcity of water in the Hongshui River Irrigation Area, because before the pilot project started, only two out of every three mu of land could get irrigated. In fact, water rights trading appeared before the start of the pilot project. After the experiment started, the institutional environment underwent significant changes in the following three aspects: (1) the issue of water withdrawal permits means realization of the initial allocation of water rights at least in form; (2) as the water rights trading during the experiments was legalized and to a certain extent, encouraged by management departments; (3) Water Users Association was set up and it could operate successfully. The above enforced institutional change increased the exclusiveness of private water rights and the quality of water for agricultural use improved, thus lowering the costs of market allocation of water rights and making water rights transfer a possibility.

The changes in the above three aspects of institutional environment does not necessarily lead to water right trading. The following background merits attention in this case: (1) water of the Heihe River Basin mainly comes from melted snow from high mountains and there is not much change in the annual water volume. In addition, the water is supplied by reservoirs in the Hongshui River Irrigation Area, so there is little uncertainty of the irrigation water. Such water resource has further lowered the costs of market reallocation of water rights; (2) the value of water in the irrigation area is very high. Although water rights transfer was prohibited before the experiments, water rights trading had long existed discreetly, forming a non-formal system of water rights trading. This system lowered the cost of institutional change. Due to the legalization of the water rights trading, the original underground trading became publicized, making it easy for the water market to develop.

Case 1 shows that water rights trading has three special features: the first one is the background of special experiments; the second is special conditions in the irrigation area; and the third is the special institutional tradition. These particularities go together to greatly lower both the dynamic transaction costs for introducing the water market and the static transaction costs of the operation of the water market. It has also been discovered that no water rights transfer took place by August 2003 in the Liyuan River Irrigation Area of Linze County, which also carried out similar experiments, issued water withdrawal permits and wrote documents on the management of water rights trading. Neither did it in the other irrigation area of Zhangye. This has, in turn, proved that the costs of the operation of water market were very high and the cost of introducing market would be even higher. So, under the current social conditions, only when the market transaction costs are significantly lowered, is it possible for water rights trading to happen in areas that greatly demand for the introduction of market mechanism.

In addition, it is highlighted that the water rights trading in the Hongshui River Irrigation Area is at a primary stage as (1) it is only one-to-one water rights transfer; (2) the object of transfer is the water volume of the period or short-term trading; (3) the current scale and amount of trading are still very small, about 1% of the water used for irrigation. It has been further discovered that even in the Hongshui River Irrigation Area that has a special background, there are still many obstacles on the way of further development, which is manifested in the feeble basis for the operation of the water rights market: (1) there is a lack of measuring facilities and the water volume is measured by the rule of thumb, far from being accurate; (2) the validity term of water rights is not definite, thus making peasants filled with anticipated misgivings about the efficiency in water saving; (3) it remains to be studied as to who should fix the prices of water rights, the market, between buyer and seller or by government departments; (4) the ways of trading and trading process remain to be explored, such as should it be conducted by the peasants water users association or privately or by announcement after trading and how to handle water rights that are not traded; and (5) studies are also needed about the operation of the peasants' water users association. All these factors in fact add to the cost of the operation of the water rights market, thus obstructing the happening of more water rights trading.

Is Case 1 representative and is it worth spreading? The author holds the view that the historical experience of the irrigation water rights trading in the Guanzhong area during the Qing Dynasty may provide reference to answering the question. The limited materials available show that the water rights trading during the Qing Dynasty once reached a very advanced stage, with water rights and land rights separated and trading being long-term transfer and the prices fixed by the market. The trading was active and happened in a vast area. In contrast, the water rights trading in the Hongshui River Irrigation Area was only too elementary.

In ancient China, water rights trading was prohibited officially. It was in such context that the market of irrigation water rights in the Guanzhong area during the Qing Dynasty reached a very high level and that explains the inherent law. Chapter 5 of this book studies the historical phenomena and holds that the changes in the scarcity of water resource were the fundamental reason for water rights trade to prevail. The unprecedented growth of population during the Qing Dynasty resulted in a huge change in the relative price of water and land resources and the water rights value, which was attached to the land right rose rapidly. If water rights could not be transferred, the arid land that used little water would face a big opportunity cost and the land would suffer from water shortages would face profit losses. The voluntary trading may improve the welfare levels of both parties for the trading. It means the increase in water resource scarcity would bring about greater demand for optimal allocation of resources and the motivation for separating water rights from land rights. The demand for putting water rights on the market would also increase. The active land trading in the Qing Dynasty stimulated the separation of water rights from land rights.

In contemporary China, however, the land held by farm households is not tradable and the rights of irrigation water are still attached to the land rights. There are not the preconditions for water rights trading, in particular long-term rights trading, at the higher level. It took a long time for water rights to separate from land rights in the ancient society. On the one hand, there is no sign of such separation in current agricultural development. On the other hand, although water resource scarcity is already very obvious in the contemporary society, there are also a series of factors inhibiting the rise of the relative value of water resources. What is fundamentally different from the ancient agricultural society is that China now is marching toward industrial society. Under such social conditions, the importance of irrigation water rights has already dropped sharply to the state managers. The decline of the dependence of peasants on agricultural production for income has also made the value of irrigation water drop sharply. Besides, trading in 'water path' in ancient times was very high in economic value due to its long-term nature. However, the water rights trade in Case 1 in the short term yields very low economic benefits to peasants. Such benefits are even limited in the short-term trading if comparing costs of operating the trading system with the cost of introducing the system.

There is another important factor associated with the benefits of water rights trading. The amount of land that Chinese farmer households held is very small in size, averaging about two mu per person and that has made the water rights attached to the land very small. Even if the water economized may be sold, the economic benefits would be extremely limited. The lower such benefits, the smaller the incentives for optimizing the allocation of water rights. This is learnt from the China Environmental and Development International Cooperation Committee that in the Murray-Darling Basin of Australia where water rights trading is very developed, each farmer has about 3000 hectares of land on average, about the total of a middle-sized township in China. Obviously, the economic benefit of Australian farmers from water rights trading is very significant. This is quite different from China. In reality, a water user in Australia is equal to a group in China and the water is transported by pressurized pipelines instead of open channels. Such technical and economic conditions are of vital importance in comparative institutional analysis. According to the materials obtained by me from its investigations, in the middle reaches of the Fraser River basin of Canada, each farmer holds 1500 hectares of land on average (Hart, 2016a, 2016b; Mitchell, Priddle, Shrubsole, Veale, & Walters, 2014; Speed, 2009a). The extreme scarcity of per capita resources is in fact an important explanatory factor behind the huge differences between the Chinese and western civilizations.

From the above factors, we can understand that there are no basic prerequisites for voluminous happening of long-term water rights trading among peasant households in contemporary China. Water resource scarcity was only one of the prerequisites for active water rights trading in the Guanzhong area during the Qing Dynasty. There are three other prerequisites behind this historical event: (1) technical changes over more than 1000 years have made water measurement more accurate; (2) the introduction of the water registration system and its long continuity has made water rights to acquire long-term stability; (3) folk rules have played an important role in maintaining the water market. The above three factors are interconnected and their common action helped improve the quality of water rights held by farming households, thus greatly lowering the costs of water rights definition and water rights trading. But the contemporary society lacks the above three prerequisites. In Case 1, the experiments in the Hongshui River Irrigation Area involved issuing water rights permits, setting up peasant water users association and introducing volumetric facilities. It would take a long period of time for these factors to reach the level in the Guangzhong area during the Qing Dynasty.

Besides, the active water rights trading in the Guangzhong area of the Qing Dynasty was also associated with the heterogeneity of water users. The theoretical part of this book points out that the greater the heterogeneity of water users are, the higher the cost of water allocation by administrative means and the lower the costs of water allocation by the market are. Generally speaking, the heterogeneity of land users is owing to the differences in the amount of land held, the differences in the characteristics of users (such as economic conditions), differences in land quality and differences in crop structure or crop planting methods. On this basis, it is understood that the heterogeneity of farming households in the Guangzhong area of the Qing Dynasty was very strong. However, in contemporary China, farmers hold their land on the equitable basis and land is not allowed to trade. The degree of economic division of farmers is fairly low. The crop structure and planting methods are very similar in given areas, which determine that the heterogeneity of farming households is very low.

The path for institutional change shows that water trading in the Guangzhong area of the Qing Dynasty was the result of natural evolution or induced change. Chapter 5 shows that the irrigation management system of ancient China underwent thousands of years of technical progress and institutional change before water rights trading appeared in the middle and late periods of the Qing Dynasty. As the water market of the Qing Dynasty was the result of induced change, the water market had good institutional compatibility. In Case 1, the change is enforced and the new institutional arrangements are obvious incompatible with the original one. The five obstacles listed above are the evidence. This problem is more outstanding in other areas. It can be imagined that even water rights trading is introduced in an enforced manner, the new institution introduced would find it hard to operate effectively due to institutional incompatibility.

The above factors imply that contemporary China does not have the prerequisites and institutional environment for irrigation water rights trading to happen in large amounts. That is determined by the high transaction costs of such institutional arrangements. The extension of market into the irrigation area tend to opt for the ways of water supply market, because the water supply market does not need defining water rights directly and therefore may economize the transaction costs of the trading of irrigation water. However, the development of water supply market in irrigation is, of necessity, falling far behind the development of water supply market in urban areas. If we say that the urban water supply at present has already developed into a pattern of 'mainly relying on the market, supplemented by planning', the water supply for irrigation is still considered as a pattern of 'mainly relying on planning, supplemented by market'. The main restricting factors in the development of irrigation supply market include the nature of irrigation channel network, conditions of volumetric facilities, and the bearing capacity of irrigation users. With the further development of the economy, the market elements in the irrigation water supply system would increase.

7.3   Water Rights Trading at Group Level

In legal sense, China's water rights are reallocated at the group level by administrative means. According to the Water Law, the water withdrawal permit is valid for five years. In 1999, all water withdrawal permits from the Yellow River issued by YRCC were reviewed. The principle for verifying water volume is laid out in water allocation scheme of the Yellow River. The withdrawal volume was verified on the basis of the comprehensive appraisal of the production development and the actual amount of water used and the water economizing measures by water withdrawing units or areas. In the current water permit system, how the new projects obtain increased water volume? Ministry of Water Resources issued in 1999 a circular on strengthening the management of water withdrawal permits (document number: *No. 520 [1999], September 14, 1999*), demanding strict examination of water withdrawing projects and tightening the issue of water permits and stopping approve big water-consuming projects in the areas suffering from severe water shortages. The water volume to be increased equals the amount saved by users through special measures or potential water saving methods. This shows that applying for additional water has to ask the administrative management departments for water. In a certain area where the total water withdrawal does not reach the ceiling, application of additional water may be satisfied; for areas whose water withdrawal has reached the ceiling has to obtain additional water from what they save. But how to solve the problem of additional water if there is insufficient water using quota obtained from water saving? There are many cases of buying irrigation water rights to solve the problem, among which the author has investigated one of the earliest cases discussed in Case 2.

Case 2: Industrial Enterprise Buying Irrigation Water Rights in the Yellow River Basin

In around 2000, the Datang Tuokeduo Power Company of Inner Mongolia, a state held large power generating enterprise, decided to launch its second phase project with four 600-mw generating units. Technical economics appraisal showed that the cost of water cooling system was much smaller than air cooling system. But the water cooling system required water. After consulting the local government, the power company was ready to invest in water-saving irrigation projects in the locality in exchange for additional water needed.

The company submitted a report to the Planning Commission of the Inner Mongolia Autonomous Region on September 21, 2001, committing to invest 89.5 million yuan (about 13 million USD) in the water saving irrigation project of five major irrigation areas and, in exchange, will get the use right of 50 million of the 5.86 billion m^3 of Yellow River water. The five irrigation areas included the Southern Bank irrigation project, the Yellow River Bend irrigation area, the National Unity Irrigation area, Dengkou water lifting irrigation area and the Madihao irrigation area. The transformation project will change the flood irrigation into irrigation through water canals, which would benefit 1.9 million mu of land and the water to be saved will reach 55.15 million m^3.

The power company was not the only one that did so. Starting from April 2003, the YRCC, the Inner Mongolia Autonomous Region and the Ningxia Hui Autonomous Region carried out experiments in water right swap, that is, regulating industrial water and irrigation water by making industrial enterprises invest in water saving projects to benefit agriculture. By the end of December 2004, YRCC had approved five such projects, three in Ningxia, and two in Inner Mongolia, with a total investment in the transformation of irrigation projects reaching 326 million yuan and the annual volume of water swapped reached 98 million m^3. In order to ensure the smooth-going of the water rights transfer experiments, the two regions all set up their own steering group headed by directors or deputy directors of the water resource bureaus of the two regions. The steering groups mapped out implementation plans. Part of the water saving projects has already started (source: interviews and notes during fieldwork).

Case 2 happened in Ningxia and Inner Mongolia. According to the Yellow River Water Allocation Scheme, Inner Mongolia and Ningxia already used up their quotas. After the YRCC tightened the total water volume to be used by riparian provinces and regions, it no longer approved applications for additional water by regions that have used or over-used their quotas. Due to the hard external constraint of the total water volume of Inner Mongolia and Ningxia, the two regions had to solve the problem of additional water needed by themselves. The normal practice when additional water is needed is that the government departments concerned have to raise water volume by tightening water management or reduce the planned water quotas. If the government management department cannot do so, the problem would be left to groups that need additional water because the existing water withdrawing projects cannot infinitely save water to satisfy the new requirements.

For new projects that need water, they have only two channels to get it: one is to ask the government department for water quotas and the other is to buy water from other water withdrawing units. Normally, the period of asking for water is long and uncertain while buying water involves big direct investments. When the cost of asking for water is higher than the cost of buying water, the new projects would have the motivation for buying water. As industrial projects have higher efficiency of using water and have the payment ability, it is very likely for industrial projects to opt for water trading. Agriculture, however, is the main water user and its efficiency is low and it has the potential for saving water and use the water saved to satisfy those that need more water.

In Case 2, we have observed that the non-performance of government management departments under the current institutional framework is the main reason for the market transfer of water rights to happen. It is because of the failure of the administrative means in regulating water rights, in order to obtain water rights groups that need additional water have to resort to market regulation. Water rights trading in Case 2 is, in essence, the result of the absence of the government. There are two possibilities to account for government absence: one is difficulty in implementing the current system and the other is the reluctance of the government to perform its duties. Whatever the reason is, they all go to show that the costs of water rights regulation by administrative means are high. What happened in Case 2 as viewed by the theoretical framework of this book is that the market reallocation is more cost-effective than administrative means, thus lowering the static transaction costs in the market reallocation of water rights. The active attitude toward such water rights transfer helped reduce the dynamic transaction costs of introducing market elements into the reallocation mechanism, making the institutional change happen smoothly.

Chapter 6 points out that under the current water management framework, the quality of water rights is low, which is favorable for reducing the costs of regulating water rights by administrative means. Case 2 implies that with the transition to the market economy, there have appeared many groups with their independent interests and their awareness of rights has been enhanced, thus making it more difficult to use administrative means to regulate water rights. On the other hand, the current water law framework provides for a set of administrative allocation systems and the implementation of the systems needs costs that would grow with the development of the market economy. The reason why administrative means becomes ineffective in regulating water rights is the high institutional costs or inadequate payment for institutional costs. From the angle of the current legal system, the government should opt for increasing payment for the costs to make administrative means to work better rather than giving up these means.

But in practice, what often happens is that the use of administrative means to regulate water rights often encroaches upon the legal rights and interests of water users (irrigation water). Since reform was carried out, an average of more than one billion m^3 of irrigation water have been occupied by industries and cities every year, in most cases free of charge. This is in fact a plunder of the rural resources.[2] Buying of irrigation water by industrial and cities in Case 2 is favorable for protecting the interests of the rural areas and should, therefore, it is considered as a progress.[3] In Case 2, in the process that water rights were converted into industrial use, the irrigation water rights are exercised by local government and that is also easy to encroach upon the interests of the rural areas where water rights quality is low.

Water rights transfer in Case 2 is achieved through coordination by the local governments and through negotiations between enterprises and the government. In this case, the right buyer is the enterprises of independent accounting that have emerged in the market-oriented reform and they are true market players. But the rights assigner is the irrigation area, which is not independent market player. This is also true in other places. The current rural water withdrawing groups are mostly public organizations or state-owned units, which are not market players in the true sense. In order to stimulate the water right transfer from agriculture to non-agricultural departments, local government exercises the rights on behalf of the rural areas that is a most cost-economizing institutional arrangement. It is regarded as a reasonable option for the present stage and also the main form of such trading in the future. The biggest challenge to such water rights transfer is how to ensure the impartiality, justice and transparency of the government in the actual operation so that the interests of farm households are better protected and the contracts are faithfully performed.

7.4   Water Rights Trading at Local Government Level

Water rights allocation at the local level is regional allocation of water. The 2002 Water Law provides that regional water allocation should be based on the basin planning, medium- and long-term planning and trans-boundary runoff allocation schemes and should be subject to total quantity control. The regional initial allocation of water rights may be regarded by administrative means. The adjustment of water rights is also conducted by administrative means. However, regional

---

[2]information collected at Water Rights Seminar organized by the Ministry of Water Resources, on August 27–28, 2001, in Beijing.

[3]In the planned economy before reform and opening up, the shortage of water governance input by the state was mainly made up for by labor days put in by the peasants. In the whole age of the planned economy, investments in water resources main came from state and the rural people, with the state investment accounting for 2/3 and that by rural people, 1/3. The labor put in by peasants may be regarded as the source and basis for irrigation water rights held by the rural areas.

allocation of water faces the same dilemma. With the progress of market-oriented reform, regions, like groups, are more and more becoming market players of independent interests. If trans-boundary runoff allocation schemes are fixed and implemented, localities would have the awareness of independent rights. The stronger such awareness is, the more difficult for the superordinates to regulate water rights. The motivation for introducing market elements would come into the reallocation mechanism. In order to examine the nature of water rights trading at the local level, we must first of all, look at China's first water rights trading that once aroused a strong repercussion (Case 3).

Case 3: Water Rights Transfer Between Dongyang and Yiwu in Zhejiang Province

Toward the end of 2000, Yiwu City in the Central Zhejiang basin bought permanent water use rights of nearly 50 million m^3 of water for 200 million yuan from its neighbor city of Dongyang. The water rights transaction opened the precedent in China's water rights system reform, according to experts from the Ministry of Water Resources.

Dongyang City is in the upper reaches of the Jinhua River while Yiwu in the lower reaches. But their respective water resources available varied greatly. Dongyang enjoyed the plentiful water resource, averaging 2126 m^3 per person. Apart from meeting its own needs, it sees more than 3000 tons of its water flowing into Dongyang. But Yiwu City has a population only about 80% that of Dongyang, but the per capita water resource is only half that of Dongyang. In recent years, with business growing fast, Yiwu City has rapidly been expanded, with its population growing to nearly 350,000. In the process of urbanization, water shortage has become a factor seriously restricting its development.

The city government called many meetings of water experts to seek counter-measures. Experts turned their eyes to Dongyang, which has two large reservoirs, one of which has a holding capacity amounting to 186% of the large and small reservoirs combined in Yiwu. As the reservoir is situated near the river source, there is no pollution. The two cities are barely 10 km apart. It would be no better thing if it could divert water from Dongyang.

But the sagacious Dongyang people have also turned their minds toward their abundant water resources. If they could transfer 1/3 of their surplus water and retain 2/3 as reserves for future development, it would not affect irrigation and urban water supply and may use the proceeds for accelerating infrastructure construction of the city. Toward the end of last year, the two cities signed a water rights transfer agreement. By the agreement, Yiwu City bought 49,999 million m3 of water use rights from the Hengjin Reservoir of Dongyang City for a lump sum of 200 million yuan while Dongyang still retains the ownership of the reservoir and is responsible for the operation and maintenance and Yiwu pays management fee for 0.1 yuan per m^3 of water supplied. The pipeline project should be planned, designed and invested by Yiwu City. The project is scheduled for operation in January 2005 (Gao, 2006; Speed, 2009b).

In Case 3, the water rights trading is taken place against the background that the two neighboring cities have water supply and demand problems. Yiwu suffers from serious water shortages due to rapid urban development while Dongyang has two large reservoirs, with plenty of water resources and there is a big potential to tap. Dongyang, through water economizing projects and new water projects, has got surplus water supply, with the cost per m^3 less than 1 yuan. When water is transferred to Yiwu, it can get 4 yuan per m^3, with a total net gain of 150 million yuan. Including the water fee income of 5 million yuan per year and profits from power generation, its gains would go far beyond 200 million yuan. On the part of Yiwu, although it has paid 4 yuan for every cubic meter of water transferred, it would cost 6 yuan per cubic meter if it builds its own reservoir. The deal saves Yiwu 100 million yuan.

Concerning the Dongyang-Yiwu water rights transfer, I visited the Zhejiang Water Resources Department of the water resources bureaus of Dongyang and Yiwu during my fieldwork in March 2001, seeing the Hengjin reservoir and its trunk canals, the Hengjin water-efficient irrigation project and the Yiwu Badu reservoir and water works of the city. My field investigation illustrates that although the two cities share the water from the same river, there is no water allocation scheme due to abundant run-off. The water resource is in a state of open access. Yiwu suffers from water shortage because the river water is polluted and water lifted is used for irrigation only. The city is eager to expand urban water supply and the use of the quality water from the Hengjin reservoirs has become the optimal option in terms of technical economics. To achieve this, under the current institutional framework, it needs to report the superordinate organs and propose to divert water from water-rich Dongyang City and then the surperordinate organs would do coordination. Almost all water diversion projects have to be done by this administrative means. However, Yiwu did not do so. Instead, it opted for buying water rights. The main reason is that the cost would be too high to use administrative means (Yahua Wang, 2001).

The biggest advantage of traditional inter-regional water regulation lies in the unconditioned benefits as the central finance or superordinate finance would pay the bill. However, such administrative means of water regulation is time consuming. Furthermore, due to lack of interest compensation, parties concerned would find it hard to reach unanimity. In Zhejiang, where the water right trading occurred, it lacks the precondition for investment by the central finance. The local irrigation projects are mainly undertaken by the locality, with very little subsidies from the provincial and central governments. The fiscal subsidy for the building of reservoirs is only 10% of the total cost. Yiwu is economically well developed and can pay the cost from its own pocket. If Yiwu had opted for reliance on the superordinate organs, the city would have to wait and it does not pay off to have the little subsidy from the superordinate finance. In such circumstances, Yiwu opted wisely for its independent solution and realized water rights turnover through market means.

The reason why Yiwu opted for buying water instead of asking the superordinate organs for water quota through administrative procedures is that the cost of buying water is far less than the cost of the administrative procedures. What Yiwu has lost is the meager subsidy (about 20 million yuan) from the superordinate level for building reservoirs. But what subsidy opportunity cost that it has got is very high, as buying water may quickly solve the water shortages in the city. For Dongyang, the diversion of water from the city by orders does not have any incentives while selling water may liquidate its water assets. As the trading benefits both parties, there is a strong desire for striking the deal.

The occurrence of water rights trading in Case 3 has its given social, economic and institutional background. Water infrastructure facilities are public goods, requiring huge investments. Yiwu is one of the top 100 cities of China in terms of economic development. Its per capita GDP in 1999 was as high as 16,000 yuan, 2.4 times that of the national average. It is strong in fiscal absorption ability and fixed assets investment. The total fiscal revenue of the city in 1999 reached 634 million yuan and its fixed assets investment came to 3.019 billion, higher than the average level in Zhejiang Province as a whole and far higher than the national average. According to the water transfer agreement, the water right transfer fee of 200 million yuan will be paid off in five years, averaging 40 million a year, accounting for only 6.3% of the city's total revenue. The whole water supply project costs 700 million yuan, averaging an annual investment of 140 million, accounting for 4.6% of its total fixed assets investment. This brings no fiscal pressure on the city; neither will there be investment squeezing-out effect, indicating the strength of Yiwu in buying water rights.

If we say that developed economy has provided the material base for water rights trading, the high development of the market and strong sense of commodity on the part of the people have created the social environment. Yiwu is China's largest small commodities distribution center while Dongyang is a city noted for its construction trade in Zhejiang. Both have edged into the top 100 counties in the country. Similar to Wenzhou, the area has flourishing industry and commerce and has a thick atmosphere of division of labor and trading. Local government officials have a strong sense of commodity. Besides, the two cities have special partnership relations and know each other well and exchange information fully, thus greatly reducing transaction costs. That is, too, an important background for the successful trading. Dongyang and Yiwu are neighbors and have cooperated well in the construction of roads and airports. They have exchanged personnel frequently. Dongyang has more than 200,000 people doing business in Yiwu. I was told that "sharing resources, making up for each other's weak points for common development. This is the common road for development of both cities." during my interviews with an officer in the Zhejiang Water Resources Department of the water resources bureaus of Dongyang and Yiwu. A leader of Dongyang City commented: "the two cities have set their eyes on a unified market instead of segmentation. They view cooperation from a strategic point of view rather than gazing at the money bag". The above analysis shows that water rights trading in Case 3 have a special economic, social and geographical background.

In Case 3, the rights traded between the two cities are not based on the rights allocated, but the de facto rights occupied by one party or de facto rights defined by natural forces. For other areas where there is no water allocation scheme between upper and lower reaches of a river, similar situation would have led to water disputes. However, in this case, as under the traditional administrative system, Yiwu is unable to effectively solve the problem by itself and it has the ability of using the market force, which has very low transaction costs, that is why the trading was done to the satisfaction of both parties. We may presume that if Yiwu does not have the economic strength to buy water rights and the transaction costs of using market are high, the outcome would have been most likely to be water disputes between the two cities. Besides, superficially, Dongyang and Yiwu are distributed in the upper and lower reaches of a river, the relations between the two cities are like a cross-basin relation due to the special reasons mentioned above, so the water allocation problem between basins does not have generality. As comparison, let us look at water rights trading between the upper reaches and lower reaches of the same river, which happened in the Haihe River Basin (Case 4).

Case 4: Cross-Province Compensatory Water Diversion in the Upper Reaches of the Zhanghe River

From May to mid-June of 2001, the Zhanghe Upper Reaches Management Bureau of the Haihe River Conservancy Commission organized five major reservoirs within the territory of Shanxi Province to provide emergency supply to major irrigation areas in Henan and Hebei provinces in the lower reaches of the Zhanghe River. The total volume of water to be supplied was 50 million m^3, at the negotiable price of 0.025 yuan per m^3. This move effectively helped the irrigation areas in the lower reaches of the river solve the irrigation problem for summer crops and ease the sharp contradictions in recent years between Henan and Hebei provinces in the lower reaches of the Zhanghe river, thus warding off water disputes on the border areas.

The Zhanghe River is originated in Shanxi and flows through the border areas between Hebei and Henan. With the economic and social development in the river basin, water shortages became very acute in recent years, especially during the irrigation peak period, which happens to be dry seasons every year. Water disputes and even fighting often occurred for water, having an adverse impact on the social stability and economic development. That caught the attention of the central leadership. In recent years, Premier Zhu Rongji and other central leaders made a number of important notes for handling water disputes.

Since the winter of 2000, sustained dry spell has gripped the North China area. Severe drought hit the Linzhou city's Red Flag Canal, Tianqiao Canal and Yuejin Irrigation area of Anyang county in Henan Province that directly draw water from the Zhanghe River and Hebei's Baiyi Irrigation area. Because of the severe drought, there was no way of planting the summer crops. The basic water level of the Zhanghe River was low and water supply was strained on both banks. In order to ease the difficulty in water use by irrigation areas and ward off water disputes that likely to happen at the turn of spring and summer, the Zhanghe Upper Reaches Management Bureau of the Heihe River Conservancy Commission mapped out a water diversion plan after investigating the water storage situation of the large and medium-sized reservoirs in Shanxi and the demand for water in Hebei and Henan and conducted consultations with the three provinces. The related units of the riparian provinces reached common understanding on cross-provincial water diversion and signed a water supply contract.

In order to ensure successful water diversion, the Zhanghe Upper Reaches Management Bureau of the Heihe River Conservancy Commission coordinated the time and flow of the water from the reservoirs in Shanxi and rationally arranged the time and volume diverted to the irrigation areas in Henan and Hebei. People were organized to strictly control the weirs and allocate water volume according to contract. Changzhi City holds that the use of water above the flood limit in the upper reaches to supply to lower reaches in a compensatory way may optimize the water resources allocation and stimulate a benign cycle of water management units. This water diversion is a useful attempt. Anyang, Linzhou and Shexian in the lower reaches that suffered enough from water shortage and water disputes bought water which ensured peace and promoted local economic development. The riparian people who have not seen so much water for several years clapped their hands for the move, calling it relief water and timely water in praise of the advantages of compensatory diversion of water in a unified manner.

An investigation shows that there are a dozen large and medium-sized reservoirs in Shanxi in the upper reaches of the Zhanghe River, with a total storage capacity of nearly 400 million m³. To optimize the water resource allocation by economic means has played a positive and effective role in easing the super and demand and settling water disputes. (Source from a report of Zhanghe Upper Reaches Management Bureau of the Heihe River Water Committee: "Collection of Materials on the Cross-Provincial Water Regulation in the Upper Reaches of the Zhanghe River", July 2001.)

Different from the long-term water rights trading between Dongyang and Yiwu, what happened in the Zhanghe River is a short-term water rights trading. The Zhanghe River is a trans-provincial border river that has caused frequent water disputes. From the 1950s, water disputes have happened frequently between Henan and Hebei provinces. In 1989, the No. 42 document of the State council mapped out a water allocation plan and the Haihe River Conservancy Commission built a series of water allocation projects. But when the water allocation plan was mapped out, the base flow of the river was reduced to around 10 m³/s. Such a flow has fallen far short of demand with the changes in natural conditions and economic and social development. In such circumstances, the Haihe River Conservancy Commission was thrown into a dilemma of no water to allocate even if there are water allocation projects. The fight for water during dry seasons has been intensified. According to the materials collected by the author, there was no cross-provincial water allocation plan among the three provinces. The water in the reservoirs in the upper reaches of the Zhanghe River is like that of the Hengjin reservoir in Dongyang, it is de facto right occupied by the upper reaches under the open access conditions. With such a water resource pattern, the lower reaches find it profitable to buy water rights to meet their own needs while the reservoirs in the upper reaches may derive some income from the water above the flood limits. That is the basis for the successful compensatory water diversion. The basin organization played the intermeriary role, thus lowering the transaction costs and ultimately making the water diversion a success.

Case 4 is a special case, in which water transfer is based not on regional plan but on the de facto rights occupied by the upper reaches. If the superordinate management department has the ability to involve in management, such de facto rights would be strongly challenged and water may be regulated among different regions by administrative means. Case 4 shows that the superordinate administrative intervention and regulation capacity is far from being adequate in face of the conflicts of interests among independent interest groups. The compensatory water rights regulation in Case 4 may be regarded as something transient, because what is transferred lacks legality. This practice of the lower reaches buying water monopolized by the upper reaches could not last long and is not worth spreading.

Now let us compare Case 4 with another emergency water diversion in 2002 in the Yellow River basin. In 2002, severe drought hit the whole of the Yellow River basin. The worst hit was Shandong Province. The YRCC, basing itself on the water allocation scheme, cut the quotas correspondingly for riparian provinces. Shandong was thrown into great difficulty in irrigation. It appealed to the central government for diverting water from the upper reaches and the appeal was approved. YRCC began to implement the scheme of diverting water from the Longyang Gorge in the upper reaches. Shandong ultimately got the irrigation for its autumn crops free of charge. The implementation of the scheme also cut part of the quotas for Inner Mongolia and Ningxia (Wang, 2003b). This practice presents a sharp contrast to that in the Zhanghe River water diversion in Case 4. The difference lies in the fact that the State Council has the water allocation plan for the Yellow River and the annual water use plan is mapped out and implemented by YRCC. The water rights of riparian provinces and regions are clear. YRCC exercises unified water diversion and monitors the water use by riparian provinces and regions, thus laying the foundation for water diversion by administrative means. In the Zhanghe River, however, the basic water management organization does not have the strength to exercise unified water diversion and there is no water allocation scheme. The basin commission does not have the ability to enforce water diversion, and it is the underlying reason for the compensatory water diversion to happen in Case 4. However, we should also notice that administrative transfer of water rights has its corresponding costs. The price of the administrative diversion of water that happened in 2002 with the Yellow River is the contempt of the areas in the middle and upper reaches of the river, whose interests are impaired. If a certain degree of market elements should be introduced as in Case 4 and the areas that had their interests lost should have been given some compensation, more satisfactory results would have been achieved.

7.5 Water Bank: Water Market at a Higher Level

'Water bank' is a market form under the regulation of superordinate decision-making entities to effectively lower transaction costs. It is like a virtual reservoir, absorbing the surplus water from users, who may withdraw water when needs arise. Its functions are like a financial bank, which absorb deposits and issue loans. As water resource is strong in mobility and has big fluctuations in hydrological characteristics, there are usually big gaps in space and time between supply and demand. The one-to-one spot trading is costly. The establishment of a water market may lower transaction costs. This form of 'water bank' is extensively adopted in western economies. For instance, in 1991–1992, California of the United States was hit by severe drought and water was in serious short supply. The local government established a water bank to buy water saved by users for use by those who suffered from serious water shortages (Israel & Lund, 1995). The establishment of water bank is preconditioned by high quality of water rights, because superior quality of water rights would naturally make water bank a good form of water market. In China, the quality of water rights is low and there has not been any instance of water bank in any form. But with the progress of the reform in the water management system, plans for establishing water market similar to water banks are in the works, which is explained in Case 5.

Case 5: Water Bank Experiments in the Works at the Lower Reaches of the Yellow River
Water diversion of the Yellow River is a massive systems engineering. Now the means for water regulation is monotonous and there is still the problem of disorderly water diversion, thus causing waste. To realize the sustainable utilization of the water resources, it is necessary to adopt a combination of administrative, economic, engineering, technical and legal means to ease the conflicts between supply and demand and make the limited water resources display their efficiency to the maximum.

Economically, related departments are planning to formulate rules on collecting additional fees for water in excess of planned quota and to set up a water supply market, to regulate water supply to order so as to give full play to the economic leverage role of the market, and also give shape to a scientific and practicable water pricing system, as well as compensatory and transfer mechanisms that well fit the market economy.

Major measures for cultivating Yellow River water market include: (1) to collect water fees at cost price or at a thin profit; (2) to levy tax on water resource used; (3) to allow prices to float and collect fixed amounts of fees for water within planned quotas and additional fees on water in excess of quotas; and (4) to experiment in the establishment of water banks in the lower reaches of the Yellow River and Henan and Shandong province may deposit their water within quotas not used up in the "bank", which may be used when needed or may be traded and transferred (source from internal materials of Water Regulation Bureau of the Yellow River Water

Conservancy Commission: "Proposals for Water Regulation of the Yellow River and Focus of Work for 2002").

The inter-provincial water allocation scheme in the Yellow River basin was introduced in 1987. After 1998, YRCC was authorized to exercise unified diversion in the whole basin and established a mechanism for implementing the water allocation scheme (Xia & Pahl-Wostl, 2012). The water rights of the riparian provinces and regions are relatively clear. The initial allocation of water rights based on the water allocation scheme has been widely honored. With the completion and use of the Xiaolangdi reservoir project, much attention has been directed to the functions of the project in diverting the run-off in the lower reaches. In this context, YRCC proposed experiments in 'water bank'. In fact, the experiments may be extended to whole basin if YRCC, who has the authority of unifying the water diversion of major reservoirs on the main river course of the Yellow River, further improves the monitoring system (Yahua Wang, 2013b). The river covers many provinces and regions and the use of water quotas is extremely uneven. The water bank may help realize the turnover of water quotas among different areas. It does not only promise great potential but also has great operability. It may display an important role in optimizing the water resources of the Yellow River basin.

During the field investigation in Zhangye, in August 2003, I conducted an investigation of the experiments in building a 'water-efficient society' in Zhangye. It is discovered that the city introduced the 'water coupon' system and that resulted in the problem of how to dispose the surplus coupons (Wang, 2003a). The Water Bureau of Zhangye City is working on a new plan to buy back the surplus coupons at a price 125% of the original and the water saved by enterprises is returned at 50% of the price. This may be regarded as the embryo form of water bank. This has also enabled us to see that the low operational costs of water market have given rise to a large amount of water rights trading. Such institutional arrangement is in great effective demand.

However, we must also see that water bank is a high level water market, built on the basis of water rights trading among different decision-making entities at the same level. If there is not the prerequisite for trading or it is impossible for large amounts of trading to take place, water bank would be tinged with idealist colors. In Case 5, the conception of water bank is feasible in practice. That means Henan and Shandong may independently buy or sell water quotas to other provinces and regions. This requires a high demand of the quality of local water rights. But the Yellow River basin still lacks the current conditions. The water rights held by provinces and regions can only be regarded as controlled quotas, far from acquiring the nature as assets. Therefore, water bank in contemporary China is only at the stage of theoretical exploration and it does not have practical operability.

7.6   Market Initial Allocation of Water Rights

All the cases discussed above fall into the category of reallocation of water rights, without touching the initial allocation. The theoretical discussions in previous chapters show that administrative means is the usual option in the initial allocation of water rights, because security fairness and social acceptability are the most concerned in terms of water allocation, which reflect the exact role of the government. It is difficult for the market to achieve these ends. But that does not necessarily mean that the initial allocation mechanism excludes market means. In such circumstances, it is possible to introduce some market elements into the initial allocation mechanism. Now let us discuss about the new line of thought of the plan for diverting water from south to north China, showing in Case 6.

Case 6: Water Allocation Planning of the South-to-North Water Diversion Project
The feasibility study of water diversion from south to north China has three characteristics: First, it has studied the necessity of the project from the angle of water resources allocation. Second, the project should be carried out in stages as the gigantic project requires big investments in all the routes, eastern, middle and western, which may save some investment and make the project better adapted to growing demand for water. Third, it uses water rights theories to direct the building and management of the project. During the period of planned economy, places often asked for quotas in excess of what was actually needed and when water arrived, they would complain that the water was too expensive to afford. The upshot is that many water diversion projects could not reach their designed capacity. The waste was huge. That is a problem universal with almost all water diversion projects. What is to be done? Take the middle route. Now all cities ask for water and equities are fixed according to the proportion of water volume asked for by each city. The fixing of equity means to buy water rights from the water diversion project. The more you ask for, the more the equity and the more capital required. But where are the funds from? It may raise the current price to the price when water is available from the diversion project. There are three advantages to do so. One is that it is conducive to economize on the use of water; another is that it may reasonably fix the water allocation quotas and capital funds; and the last is that the water price may experience smooth transition after the water is available from the diversion project. After hearing the report, the Premier accepted the proposal for establishing a water diversion fund. This is a very important policy decision, which may pioneer a new road for the utilization of water resources and give shape to a new system featuring macro adjustment by the state, market operation by companies and users participation in management. After water is available, dual price system will be introduced, that is, capacity price and volumetric price. By capacity price, it means payment for water rights bought regardless of the rights used or not. Volumetric price means to pay according to volume of water used. Such institutional arrangement may achieve the full use of water and display the anticipated efficiency of the water diversion project. A water system in the eastern route

has already taken shape in Jiangsu Province, which cannot only be used for diverting water but also for draining water. It also has the function of diverting flood water and waterway shipping. The water system is complicated and Jiangsu may set up a limited liability company. A few days ago, the Northern Jiangsu Water Supply Bureau already hang out its shingle. It will operate like a company. Up north in Shandong province, there may be a state-held Eastern Route Water Supply Company limited formed by three provinces, including Tianjin. The Jiangsu Water Supply Company and the Eastern Route Water Supply Company are in buying and selling relations and supply water according to agreements or contracts. This is a market under the regulation by the state (Wang, 2000).

Making cities along the water diversion routes to share part of the capital funds in proportion according to the water volume they require is, in fact, the introduction of a certain degree of market mechanism in the initial allocation of water rights. The reform has overcome the drawbacks of low efficiency in the traditional water diversion projects and may lower the costs of the use of allocation mechanism. Traditional inter-basin water diversion used to be done by the state with money from its own pocket and localities benefited from it unconditionally. In this way, the project capacity used to be too big, resulting in heavy financial burdens of the state and water resource waste. Such waste directly exacerbated the water shortage and pollution as water becomes scarcer. The advantage of tying investment to water rights is to enable decision-making entities to reflect their demand objectively and raise the utilization rate of funds and water resources.

The water diversion project is artificial, with the increase water supply capacity (increment water rights) to the northern part of the country mainly comes from project investment, thus making it different from the allocation of natural runoff. The super large project requires huge investments. If the central and local governments invest in the projects, it would entail heavy burdens on the finance. Motivated by the desire to lighten the financial burden, the state decided to seek other channels of investment by tying investment to water supply.[4] According to the current water diversion scheme, funds mainly come from payment by users and additional prices for water used in excess of quotas. Banks paid the advance, which will be repaid in installments from funds to be collected from water prices.

Case 6 illustrate that increment water rights are quite different from the initial allocation of natural runoff. The allocation of natural water resources involves a series of complicated principles, such as respecting history, domestic use first, grain security first and water source first. As the principles of allocation are favorable to different interest groups, it is very difficult to seek an allocation principle that is accepted by all groups. Although the increment water rights allocation also takes into consideration of a series of allocation principles, it is mainly obtained from the

---

[4]In fact, in ancient society, some small irrigation channels were built by peasants by pooling funds and water rights were therefore owned by collectives. There are some other small irrigation projects were run by merchants and water rights were held by private persons. All these are instances for tying investment to initial allocation of water rights.

project investment. As a matter of fact, the allocation principle of sharing water rights by making investment is readily acceptable. Under the condition of market economy, localities have increased their sense of independent rights and investors of water projects have been diversified, thus making the practice of linking water supply to investment a reasonable option in the initial allocation of increment water rights.

The initial allocation of water rights in the south–north water diversion program has introduced a certain measure of market mechanism. But it is still not market allocation. In reality, the project, dominated by the central authorities and with investment coming from the central finance, is a major strategic arrangement for the development of the national economy and it will mainly resolve the water resource shortages in the Yellow-Huaihe-Haihe basins. The purpose of introducing market mechanism is to rationally regulate the demand for water and raise the efficiency of the utilization of resources. As the local governments along the routes are not market players, it is impossible to have the incentive to make the policy of sharing investments entirely based on economic principles and furthermore, the capabilities of the areas along the routes to bear the economic burdens vary, and the differences in water resources conditions are also great. The principle of initial allocation of rights according to the size of investment is limited in the ultimate use of the water rights by various areas along the routes. To be more accurate, the initial allocation of water rights in the south–north water diversion project is a combination of administrative and market means. Besides, the water diverted will mainly be used in cities and industries and that has made it easy to raise water prices and leave it to the market force. It is very similar to urban water supply, but different only in scale, level and process. It is anticipated that the market to be formed resulting from the south–north water diversion project will be, in essence, a water supply super-market, with the basic operational mode being one of water supply between superordinate and subordinate decision-making entities rather than water market, that is, the rights trading between equal decision-making entities, which is the main concern of this book.

7.7 Summary and Comments of the Case Studies

In this chapter, we have discussed six water market cases in contemporary China and divided them into five major categories. The first four cases are the focus of the study. They are the attempts to transfer water rights by introducing market elements at the user, group and local levels. Analysis of this chapter shows that the first four cases are very unique in nature. All the cases happened in an environment in which there were many factors that can greatly lower the cost of market and the administrative adjustment of water rights was costly. It is exactly these factors that have made the introduction of market elements a possibility. In addition, another important factor is the emancipation of the mind. It is discovered that from about 2000, water rights trading began to increase. Such sudden burst of water rights trading instances was the result of the new line of thought in water governance advanced by the Ministry of Water Resources. Through emancipating the mind and theoretical guidance, the cost of institutional change has been lowered, thus making water market appear in advance in some special areas and in some special circumstances (Yahua Wang, 2013a).

As an outcome of pilot projects, Case 1 is an example of short-term water rights trading at the user level, which originated to a large extent from the special environment of the irrigation area. Compared with the trading of irrigation water rights in the Guanzhong area during the Qing Dynasty, the water rights trading in Case 1 is basic. Even so, the irrigation area does not have the necessary conditions for long-term transfer of the rights of irrigation water. The Case shows that the water rights trading among farm households in the current institutional environment will face very high transaction costs and this determines that such way of water rights trading will not take place in large amounts. However, the development trend shows that the reallocation of irrigation water rights will change toward the direction of water supply market between irrigation management departments and farm households rather than among farm households.

Case 2 is an example of long-term water rights trading at the group level, originated directly from the strict implementation of the control over the total amount of regional water use. The situation forced the seeking of new ways for settling new water demand in the area while the current administrative means was incapable of promoting water saving to increase water supply and industries were forced to invest in water-efficient irrigation projects in exchange for water rights held by agriculture. The case has explored through a feasible way for water resources to move from agriculture to non-agriculture, which has ensured the interests of farmers, agriculture and rural industry and solve the problem of increasing demand for water in economic development. This is what should be encouraged. The case also implies that, with the development of the market economy, costs of administrative adjustment of water rights are rising and those by the market are falling.

Case 3 and Case 4 are examples of water rights trading at the local government level, with the former being long-term trading and the latter, short-term trading. These two examples are trading among local governments, as the rights are occupied de facto by one party rather than that based on the already legalized rights. The special backdrop lowered the transaction costs, thus resulting in the compensatory turnover of water rights among localities, which are clearly depicted in Case 3. Although the water rights trading among local governments did not have the prerequisite of legality that should not be encouraged in practice, what are the motivations of the occurrence g of the two cases provokes deep thought. The power delegation concomitant with the market-oriented reform has greatly intensified the identity of local governments with their own interests, thus raising the costs of the superordinate government to control the subordinate government, however, the current institutions have failed to make any effective response to this. That has made the seemingly improper water rights transfer a rational option due to the absence of the government.

The case studies hint at the prospects for the development of China's water market. Firstly, at the user level, the general trend in water rights adjustment is toward water supply market rather than water rights market, which is determined by the high transaction costs of the rights turnover among users. Secondly, at the group level, there is some space for development in the water withdrawal right market, especially industries and cities, which may use investment in water-efficient irrigation projects in exchange for the use of water for agriculture to realize the transfer of water rights from agriculture to non-agriculture areas, which promises broad prospects. Thirdly, at the local government level, the pressing task at the present is to define regional water use rights according to law and regional water rights trading should not be encouraged or promoted as it does not have the prerequisites.

Moreover, successful water markets development is attributed to both long-term and short-term constraints. Long-term constraints on the development of water markets include: (1) political and social structure dominated by administrative control, which naturally excludes negotiations and trading among equality entities. This, plus the slow social development of citizens, has deprived water trading at group and local levels of political and organizational basis; (2) China is slow in the reform of property rights and it has not done enough to protect private property. The reform of property rights of such natural resources as land is even more lagging. Water resource is the most complicated natural resource and the speed of making its property rights clearer could not exceed that for other economic resources and natural resources; (3) China's meteorological and geographical conditions are unique, with big changes in precipitation and water flows. Drought and flooding are frequent. Such high uncertainty naturally excludes market and is high dependent on administrative means. The development level of water market is closely associated with the certain stages of social development. The short-term constraints for developing water markets are shown as: (1) economic transition has left complicated property right problem, making water rights reform subject to the reform of the property rights of water resources projects; (2) the development stage has made it impossible to establish enforcement and assurance mechanisms of water rights and that determines that making water rights clearer is a long process; and (3) slow progress in the water price reform in cities and towns and the administrative monopoly of water supply market have reduced the motive power of the urban water supply market to the development of other markets.

The case studies unveil the special characteristics of water rights trading. All these cases happened due to special environments that greatly lower the costs of market adjustment of water rights. When viewed from another angle, in most circumstances at present, administrative adjustments of water rights remain an option with more economical transaction costs. It has also proved that the water rights quality is generally very low at present. All these aspects are compatible with the hierarchical water rights structure, which is naturally excluding market mechanism and thus holds up the development of water markets.

On the other hand, with the deep-going market-oriented reform, market mechanism would penetrate into the allocation of all resources. Although water resource has its own particular features, naturally dependent on administrative means, the expansionary nature of market mechanism will inevitably exert pressure to bear upon water resource allocation and induce it to have the motivation of introducing market elements. This manifests itself in the following three interconnected aspects, that is the market economy requires all economic and social activities to reflect truly the prices of factors. For instance, water price must reflect its true costs; investment in non-state irrigation projects requires returns on capital; market-oriented reform has made independent interests groups diversified, thus increasing the heterogeneity of right holders and boosting the costs of administrative reallocation of water rights; the establishment and improvement of market environment have rapidly reduced the transaction costs in the allocation of all resources and correspondingly reduced the costs of market reallocation of water rights.

The hierarchy structure's exclusiveness of market and the motivation for introducing market elements resulting from the market-oriented reform have thrown the current water rights structure into a dilemma. The study in this chapter has further revealed the underlying implications of this dilemma. The market-oriented economic reform and the changes in the external environment have made it increasingly costly to manage this hierarchical governance structure. The inadequate input in the management cost by the state has caused serious 'absence of the government' in the allocation of water resources. In such circumstances, introduction of a certain degree of market mechanism is a kind of cooperation costs paid by the society, which is a compensation for the inadequate management costs paid by the government. This view goes along logically with the explanation of water crisis described in Chap. 6, that is, the 'absence of the government' due to inadequate input into management costs will make the original hierarchy structure unable to last or be kept stable. That is the institutional essence of the water crisis. The introduction of water market may, therefore, be regarded as self-adjustments of the hierarchy structure in order to maintain its stability. Water market is the product of the instability of the hierarchy structure.

In fact, 'absence of the government' behind the water crisis and water market is, in essence, the inadequacy of collective action capabilities. The collective actions over the past more than 2000 years have mainly been directed by the government. It is doubtlessly an option for minimizing transaction costs in a traditional agricultural society. Under the modern market conditions that have never been experienced, traditional collective actions, dominated by the government, have made the transaction costs too high for the society to bear, thus resulting in the extensive 'absence of the government'. As the new model of collective actions (broad participation by citizens) has not been established, it is unable to make up for the shortages of management costs resulting from the 'absence of the government'. China's current collective action have become serious inadequate during the transitional period. The dilemma in water governance in contemporary China is, in fact, the 'dilemma of collective action'. In this case, the challenge to the water governance structure is, in reality, also a challenge to the state governance structure.

Chapter 8
Conclusions and Outlook

This chapter summarizes with basic theoretical and empirical conclusions of the study concerning three issues: (1) the precise implications of water rights; (2) how to understand the problem of water rights allocation (initial water rights definition); (3) how to understand the relations between the government and market in water resource management in the economic and social transitional period of China. This is followed by the basic conclusions of the empirical study, which sum up the changing patterns of China's water rights structure in three historical periods: ancient times, from the founding of the P.R. China to present and the future. Then this chapter puts forward some specific recommendations based on the theoretical framework and the conclusions of the study. From the strategic perspective, this chapter proposes a shift from project-oriented construction to institution-focused construction. Regarding policy, it advances nine categories of mechanisms for improving water resource management system, including specific policy recommendations for each developmental stage. The book ends with the discussion about the challenge to the water governance in contemporary China in relation to the state governance.

8.1 Conclusions of the Study

8.1.1 Implications of Water Rights

Water rights are property rights to water resources, that is, rights to property in the allocation and utilization of a certain amount of scarce water resources. Due to its fluidity, recyclability, renewability and the nature as public goods and other natural and social characters, the issue of property rights is far more complicated than the

© Springer Nature Singapore Pte Ltd. 2018
Y. Wang, *Assessing Water Rights in China*, Water Resources Development and Management, DOI 10.1007/978-981-10-5083-1_8

property rights to ordinary economic assets. Holders of water rights could be a state, certain groups or individuals. Normally the state holds the rights of disposing of the water resources; groups hold water withdrawal rights and the rights to allocating water available; the ultimate users hold the rights to use. What the people are most concerned about are the rights held by groups or users, which are often attenuated due to state intervention or third party encroachment. If such rights are severely attenuated, the quality of water rights held by groups or individuals is very low and vice versa. Rights have the characters of exclusivity, duration, transferability and divisibility. Such water rights are close to pure private property rights or exclusive group property.

The low quality of water rights held by groups and individuals are often defined legally as 'administratively licensed rights (prerogative rights)', which are subject to the control by the state that holds the political rights. They are allocated and reallocated mainly by administrative means. The state imposes many restrictions on such rights and on the exercise of such rights and carries out administrative intervention. With the rise in quality, water rights are changing from administratively licensed rights to usufructuary rights (prerogative rights in rem) in the legal sense. If such rights have a higher quality due to strong protection from the state, they may even get closer to absolute property rights that are similar to the property rights of ordinary assets. State restrictions and intervention would be attenuated in this process and water rights would absorb more and more elements dominated by other economic interests, and they will be further reallocated by market means. The corresponding relations between water rights quality and other aspects of the water rights regime are shown in Fig. 8.1.

Water rights quality is socially endogenous. In usual circumstances, the bigger the challenge the society faces by water resources, the more centralized the power is and the more the state tends to directly allocate water in order to ensure water security and impartiality in allocation. This strong authority has maintained the

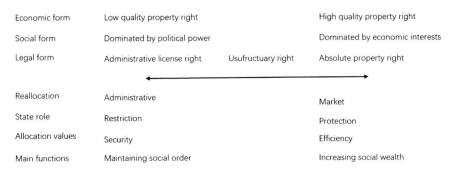

Fig. 8.1 Incremental continuum of water rights regime

order in the utilization of water resources. Consequently, the water rights held by groups and end users are low in quality. If a society does not have much challenge from water resources and water security is highly secure, the society tends to treat water resources as an ordinary economic asset and seek higher efficiency in the allocation of water in order to display its function of creating more social wealth. In such circumstances, due to increased state protection and high quality of water rights, it would be much easier for the market to reallocate water rights, and complicated water markets would develop on the basis of private property rights.

China is a society with a tradition of highly centralized power, which was necessitated from the grave challenges by water in the early period of the civilization, hence called by western scholars as a hydraulic society. This study asserts that the hierarchy structure is a response to the demand for centralized water governance system in the early period of the civilization. Under the hierarchical governance structure, the state had infinite powers in allocation of water rights and the water rights held by the private are subordinate to those of the state. In ancient times, in the name of public ownership of water rights, the political power of the state over water was prioritized and the private benefit from water use was nested in the public interests. The usufructuary rights held by individuals had long been attenuated and the water rights in the sense of assets had long been ignored. In most of the historical periods, the quality of water rights held by individuals was low, with the degree varying during different historical periods. Examination of the changes in the ancient water rights institutions shows that water rights quality improved step by step from the Qin-Han to Ming-Qing periods until a large amount of water rights trading appeared in the middle and late periods of the Qing Dynasty, when the rights of irrigation water were close to private property of farm households. The evolution of such characteristics of the rights of irrigation water in ancient times is a proof of the laws shown in Fig. 8.1.[1]

In the period of Chinese planned economy, under the framework of public ownership, the state played an important role in constructing irrigation projects and developing water resources. The state power was in the absolute dominant position in water resource management. The quality of rights of irrigation water held by farm households was very low. Since the country introduced the Reform and Opening-Up Policy, the water scarcity increased and the country began to establish a system of administrative allocation and planned use of water and clearly defined the control quotas for water to be used by certain areas, groups and individuals. Furthermore, the rise in the relative price of water, growing competition for water, and the delegation of power demand granting clear water rights to regions, groups and individuals in order to raise water use efficiency, attract

---

[1]In reality, there were no groups independent of the state in ancient China. The group discussed in previous chapters are, in fact, still part or extension of the state power. This is determined by the state governance structure.

investment in water resource development and settle water disputes. This hastened the birth of the reform guideline of separating use right from ownership, and the state still retains the ownership of water resources while granting the use rights to users. The essence of the guideline is to recognize to a certain extent the legal rights and interests of water users while stressing the state's political power over water resources. This idea touched off a great debate over water rights and water market around 2000. Water rights thus became a hot subject of talk in the whole society. The revised water law that came into effect starting from 2002 clarifies that the rights endowed by water license are water withdrawal rights. However, such rights are, in essence, still 'administratively licensed rights'. At present, policymaking departments are actively promoting the transfer of water rights (Chen, Wang, & Zhu, 2014), and this may be regarded as an effort to change the administratively licensed rights to usufructuary rights. The process of reform mentioned above reflects the slow progress in the improvement of water rights quality.

8.1.2   Theoretical View on Water Rights Initial Allocation

Water rights initial allocation is termed as the initial definition of water rights. It is regarded by policy makers as crucial to the reform of water rights in China. They hold such a view that most of the water resources in China have not been allocated, and it is essential to define clearly the initial water rights for all water users. In this book the 'definition of initial water rights' is not used, because it is easy to mislead into the following two traps: on the one hand, this expression could lead inappropriate understanding that once the initial water rights are defined, water rights would be transferred through the market based on the initial water rights; on the other hand, the way of expression carries implication that there are no initial water rights and therefore a scheme of defining initial water rights is needed. The misunderstanding could make the complexity of water rights allocation underestimated.

In fact, when viewed from the angle of economics, initial water rights always exist, whether or not water rights are allocated legally or how they are allocated. The allocation of the property rights to water resources is an objective reality. According to the framework of this study, the government always holds the rights of allocation and groups always hold the water withdrawal rights and users hold the use rights. Even if the water resources are in the state of open access and the government has given up the allocation rights, groups still hold the de facto water withdrawal rights and users still hold the de facto use rights. Without clear legal provisions, these rights are de facto rights. When unchallenged, the de facto rights, as a factor influencing acts, are just like de jure rights. Only when de facto rights are challenged, the de facto rights would reveal its differences from the de jure rights.

As the property rights of water resources are always de facto allocated, the rights held by current decision-making entities are initial water rights. The so-called definition of initial water rights is in fact an adjustment of the current water rights. So the definition of initial water rights, if more accurately expressed, is definition of water rights or making water rights clearer.

On the other hand, the definition of water rights cannot be accomplished at one go. According to the study in this book, the fundamental difference in water rights in different circumstances (such as when there is the property rights law and when there is not) lie not in whether or not the decision-making entities hold water rights, but in the differences of the characteristics of water rights they hold. This book tries to gauge the characteristics of water rights from four dimensions: exclusivity, duration, transferability and divisibility, with exclusivity being the main vector, and calls the result of the characteristics quality of water rights. The definition of water rights is, in reality, the improvement of the quality of water rights. Water rights are, therefore, constantly defined or say that water rights are marginally defined. The real meaning of water rights definition is to constantly improve the quality of water rights held by decision-making entities (including local governments, groups and users), from low quality to high quality in the economic sense and from administratively licensed rights to usufructuary rights and even closer to high quality absolute rights in rem in the legal sense. Only when the quality of water rights is raised to a certain level, is it possible to make market transfer of water rights a reality. This shows that it is wrong to think that once initial water rights are allocated and market turnover of water rights is allowed, water rights trading would happen automatically.

The theoretical study further demonstrates that definition of water rights is necessary at multiple levels with multiple dimensions. By multiple levels, it means that there are institutional options at every level between the central and local governments, between local governments and groups and between groups and users. By multiple dimensions, it means that there should be many systems for defining water rights. This study has given the three most important ones: entitlement system, initial allocation system and reallocation mechanism, which are later on expanded into three categories: allocation system, enforcement system and sustentation system. The process of defining water rights is to introduce these systems into each level and bring them to perfection. The sequential order of definition is from down-up as water scarcity expands in scope. Besides, water rights definition has multiple paths. This book has relatively subdivided the three categories of water rights regimes: the entitlement system is subdivided into input quota and resource quota entitlements; the initial allocation mechanism, and reallocation mechanism are all subdivided into market allocation and administrative allocation. There are two different paths for institutional changes when water rights are further defined in all the systems. Even so, the institutional dichotomous classification is only relative, because in practice the two institutions are mutually complementary and change incrementally from one to the other. This shows that there is a diversity of water rights definition.

The degree of water rights definition is determined by the costs and benefits of definition. If the marginal benefits of water rights definition are bigger than marginal costs, it would become lucrative to further define the water rights. Costs of water rights definition usually include (1) allocation cost, to set up and improve the allocation system in different circumstances; (2) enforcement cost, to oversee and execute the allocation system, possibly including the setting up of water market system; and (3) sustentation cost, water measuring, monitoring, statistics, reports and mediation of water disputes. There are also non-economic costs. The rise in the degree of water rights definition means the drop of the government's capabilities to coordinate and consolidate social resources, possibly threatening water security. Benefits of water rights definition by and large include: (1) to provide decision-making entities with the functions of constraints and incentives to raise the efficiency of water use; (2) if water market is introduced, the effective of optimal allocation of water resources is more significant; (3) to attract investment in water resources development to lighten the burdens on the state; and (4) in some circumstances, clearer water rights are conducive to mediating water disputes.

The fluidity, recyclability and hydrologic uncertainty have made completely clear definition of water rights (making it absolute rights in rem) very costly, far costly than that of ordinary economic assets and also than exclusive natural resources such as land. This determines that completely clear definition of water rights does not pay off. The definition of water rights in reality is always at an equilibrium level or say there is a certain degree of obscurity. The natural geographical and social and economic conditions vary from basin to basin and so are the cost-effective functions of water rights definition. That explains why there is a diversity of water rights regimes in various river basins. This book has examined in detail the inherent logic in the definition by all regimes. The options between input quota and resource quota entitlements and the options between administrative and market allocations in the initial allocation and reallocation mechanisms all follow the principle of minimizing transaction costs (static transaction costs). This logic explains the shade of differences of water rights regimes in various areas.

Water rights definition is a continuous process of institutional change. Just as natural, economic and social conditions are in a dynamically changing state, so are the cost-effective functions of water rights definition. That breaks the original equilibrium to give rise to the motivation for a change toward a new equilibrium. The increase in resource scarcity, the growth of population, technical progress, change in institutional demand or changes in other institutional arrangements are likely to become a motive force for institutional change. The rise in water scarcity is the most fundamental motivation for institutional change. Institutional change per se needs costs (dynamic transaction costs). If such costs are too high, it would obstruct the introduction of more efficient institutions. Such path dependency of institutional change may explain why the progress of introducing water rights market is so difficult in contemporary China (Coggan, Buitelaar, Whitten, & Bennett, 2013; Marshall, 2013).

8.1.3 Relations Between Government and Market in Water Resource Management During the Transitional Period

How to coordinate the relations between the government and market in the new market economy conditions is a central concern in the changes of water governance in China. Through a large amount of theoretical and empirical studies, this book has discovered that the issue is far more complicated than what used to be thought. The author views that the complexity and difficulty lies in the fact that the water resource management system has been thrown into a dilemma during the special period of transition from the old system to the new one, when it hardly had the traditional system (marked by centralization and planning) been completely established before the pressure of transition to a modern system (marked by decentralization and market) began to mount. In the following, we shall use the terminology of this book to expound on the views above and then comment on other views in existence and the line of thought for the current reform.

From the Qin empire when the first unified polity took shape, a unique hierarchy state governance structure prevailed, replacing political trading with administrative control. This was reflected in the governance of all public services, including water resource management. This hierarchical structure has kept its continuity over the past more than 2000 years. Even after New China was founded, the strong inertia of civilization has still displayed its roles and the state governance is fundamentally the continuity of the hierarchical structure. The past half century since the founding of New China is a history of the establishment, development and improvement of the water rights hierarchical structure. Large scaled institutional building did not start until the 1980s, when an administrative management framework for water resources has been established rapidly and the process has been further accelerated over the past fifteen years, with administrative water allocation mechanism established at all levels of the water rights structure, forming a complete planned allocation system or a complete hierarchy at least in form. The state policymakers are still pushing this planned allocation system toward perfection.

If it were in an ancient society, this process would continue, just as the logic established from the very beginning of the ancient empire where a series of institutions aimed at intensifying the state authority. Consequently, the state bureaucratic system would develop towards perfect; ideology would be under effective control; the state management cost would drop rapidly to a level to maintain this hierarchy governance structure; and the induced institutional change with the passing of the time would further reduce the operational costs of this structure. Now China is operating under entirely new historical conditions. The most significant difference lies in the rapid industrialization process, which is completely different from the ancient agricultural society. China is in a period of market-oriented reform that has never been seen in history. New historical conditions require new governance structures. In this context, in terms of water governance is, intensifying the hierarchical structure comes into conflict with the demand for lowering management costs. In fact, the direct effect of market-oriented reform raises the management costs in the hierarchical structure. The delegation of power and the yielding of interests by the central authorities to localities, for instance, have made localities acquire more independent interests and the costs of the state control over its local agents have been raised greatly. The diversity of main economic players and property rights owners makes the state authorities and social control power less dominant. The diversity of the sources of public investment, including water resource investment, demands the state to grant and ensure non-state rights. The diversity of economic structures and ways of production makes the ancient egalitarian distribution principles hardly sustain. In a word, the centralization of power that is required in water governance is in sharp conflicts with the daily growing demand for power sharing.

The direct outcome of the conflicts is the inadequacy of management costs required in maintaining the hierarchical water rights structure or extensive 'government absence' in practice. The state does not have the capacity to provide all public services and it can only commit its limited resources to coping with the immediate crisis and carry out only crisis-driven reforms passively, unable to continue supplying large amounts of effective management systems and maintain the stability of the hierarchical water governance structure. The further outcome of inadequate payment for management costs is the inability of continuing the hierarchical structure, the lowering of the degree of the governance hierarchy and high instability in the governance structure. That is the institutional essence of 'water crisis' in contemporary China. The emergence of water rights market is a concomitant with the instability of the governance structure, a 'by-product' of the 'absence of the government' in water resource management and a kind of compensation for the inadequate payment for the state management costs. If water rights market is extensively introduced, it would lessen the levels of the governance structure, and the hierarchy governance structure would disintegrate. That is the inevitable result of the reform. In order to maintain the immediate order, the state would do as much as possible to maintain a hierarchy governance structure at least in a short period of time of being, though such structure is flawed. In the reform process, many economic areas have to experience a transitional period or 'a period of delivery pain', so does the transition of the water resource allocation regime (Dou & Wang, 2017). Only that new regime will come much later than those in other economic areas due to its special features. The water resource management is now in such a period that the storm of transition has not arrived, but it is certain of coming, sooner or later. Thus it is necessary to get well prepared. The emergence of water rights and water market reflects exactly such efforts in the reform.

Further, it is clear that the great debate on water rights and water market differ only in the angles of approach and in the value orientation. First of all, let us look at the market skepticism. This view correctly reflects the dilemma in the current system and it upholds the maintenance of the current administrative management system. Only by taking further steps to improve the administrative management system, is it possible to eliminate step by step the de facto water rights trading which is reasonable but not lawful. In terms of market base theory, it implies the advocacy for a fundamental change of the current administrative management system and a shift of the base for water allocation from administrative means to market means, so as to meet the requirements of the market economy. The limited market theory is more pragmatic than the market base theory, stressing the realistic operability during the period of transition. It holds that the current market means may make up for what lacks in the administrative allocation system and market elements should be partially introduced to ease the current water crisis.

Market skeptics believes that water resources management under the current political and social conditions must depend on a hierarchy governance structure. This is correct, however, the problem lies in what if we are unable to pay for the costs of maintaining the hierarchical structure. So the introduction of certain elements of market mechanism should be a pragmatic option. This is the view of the limited market theory. As time goes on, the market system will gradually improve; the political and social structures will change too. So the water resource management hierarchy will become unnecessary. In this case, a flat governance structure will meet the requirements of the new social conditions. If it is such a case, it is entirely possible for the market to display its dominating role in water allocation. This may become the kingpin for the market base theory. Even for many of the current policy makers, the three representative views may reach unanimity in practice. The policy makers may be market sceptics, but in policy and practice, they may hold an active attitude toward exploration. The limited market theory is useful in action, and the market-based theory has provided policy makers with prospective objectives of reform.

In actual reform, more policy makers hold the view of 'limited market theory', regarding the market transfer of water rights as a new subject and new task in the reform of water resource management. The representative schemes put forward by policy makers are to combine total quantity control with quota management and to establish two sets of quotas. One is the total quantity control quota, that is, to quantify the water quotas and decompose them level by level right down to every basin, every area, every city, and every unit, and control quota exists at each level. The other set is micro quotas, that is to fix water use quotas in line with the total amount available, according to specific industrial products, population and irrigation areas. The two sets of quotas are applied simultaneously to achieve control in total amount and in specific quotas. At the same time, water quotas saved by users and units may be transferred on the pay basis and water used in excess of the authorized quotas has to be paid for. In this way, both buyers and sellers would try to save water, thus mobilizing the zeal for water conservation. At present, policy making departments have initially drafted a document on the management of water rights transfer along this line, which has initially fixed the transferable water rights as the amount of water saved through conservation measures and water resource protection (Wang, 2000).

What is studied above is a realistic option in the process of reform at the current stage. The emphasis is to improve the current administrative water allocation system and the market elements are only an instrument to serve the purpose. This is because, in the current social environment, the costs of water rights market operation are still very high, which is higher than the costs of the change of the reallocation system from administrative means to market means. It is, therefore, very difficult to introduce market elements on a large scale in a short period of time. The current administrative management system is still efficient and there are no alternatives in coping with the current water crisis. Thus, what can be done before the completion of the transition from the old system to the new one is to improve the current administrative allocation, with limited market elements to be introduced as a supplement, mainly in the water withdrawal permits market. In the water allocation mechanism, administrative means is still in the absolute dominant position while market means is in a subordinate position, a pattern described as 'having planning as the main means and market as a supplement'. Water market is mainly used as a subsidiary tool in the planned allocation of water quotas. This assertion is quite different from what many people have perceived, because people are accustomed to thinking that water market is water supply market but not the water rights market that is concerned by this book. The water supply market has broad prospects of development with the expansion of water supply in cities and towns with the increase in the water resource controllability. This is the most important manifestation of the expansion of the market economy in the area of water allocation. Looking into the future, with the establishment of the market economy in every aspect of the economy, the costs of administrative water allocation would grow and the costs of water rights market would drop steadily. Therefore, water rights market, mainly water withdrawal right market, would have a certain space for displaying its roles. No matter to what extent the water rights market will develop, the role of market mechanism would be manifested in the water supply market in the future.

8.2   Development Trend of China's Water Rights Structure

8.2.1   Water Right Structural Change in Ancient Society

In the ancient agricultural society, irrigation water was the main subject of allocation. Under the framework of the hierarchy water governance structure, the changes of water rights structure are mainly shown in the changes of the irrigation water rights allocation mechanism at the group-to-user level. In the more than 2000 years' history, irrigation management pattern remained basically unchanged. Running throughout the ancient society was the allocation principle of equitable use of water, with water quotas fixed according the land that needed irrigation. But still there were some gradual changes with the changes of dynasties in the long history of China. In general, the state dominated the construction of large irrigation projects during the Qin-Han period when the irrigation water rights allocation mechanism was just established. The state had a significant role to play in water conservancy management in the Tang-Song period, when the water allocation mechanism was highly developed. The main functions of the state were shifted to macro control in the Ming-Qing period, when the water allocation mechanism became mature. According to the study by this author, the ancient water rights structural change from the Tang-Song to the Ming-Qing period is concluded in the following four aspects: first, the state force gradually phased out of the micro management of irrigation; second, the entitlement system of irrigation water was shifted from input quota to resource quota; third, the use of water registration system replaced the application system in the initial allocation of irrigation water rights; and fourth, more and more market elements were introduced into the reallocation of irrigation water rights.

From the Tang-Song to the Ming-Qing period, the state gradually withdrew from irrigation management and that provokes much thought. It revealed the inherent demand for economizing on management costs in a hierarchy governance structure, which may be realized by relying on the self-governmental forces. Such change was the cumulative results of a series of induced institutional changes in the long history rather than the result of delegating powers. The explanation by the hierarchy theory of the irrigation water entitlement, initial allocation mechanism and reallocation mechanism also revealed the profound endogeneity of the institutional options and its dependence on economic and technical conditions and social environment. What has special academic value is the theoretical explanation by water rights trading in the middle and late periods of the Qing Dynasty. The author holds the view that this is the result of rising costs of administrative adjustments of water rights and the lowering of the costs of market reallocation. What is more interesting is that no ancient Chinese authorities had ever recognized the legality of water rights trading. The large scaled water rights trading in the period is, in essence, a reflection of the dilemma faced by the state in management, which is the conflict between equitable allocation by political will and the pursuit of economic interests through market means. Now the same situation is a cause for concern of the Chinese authorities. The cyclic changes between administrative and market reallocation of water rights as revealed by this book serve as a reference for coordinating the relations between the government and market in water management in contemporary China.

The evolution of water rights structure in ancient times has corroborated the incremental change continuum of the water rights regimes shown in Fig. 8.1. From the Qing-Han to the Ming-Qing period, with the gradual rise in the quality of irrigation water rights, other aspects of the water rights regime, such as functions of the state and ways of reallocation, all move correspondingly to the right or left in the continuum. The empirical study of the changes in the ancient water rights structure shows that the rise in the quality of irrigation water rights is a contributing factor to the large scaled irrigation water rights trading in the middle and late periods of the Qing Dynasty and also the cumulative results of a series of technical progress and institutional changes after Tang-Song. This shows that the rise in the quality of water rights is the outcome of long-term institutional evolution rather than momentary accomplishment.

8.2.2   Changes of Water Rights Structure Since the Founding of New China

The water rights structure at the group-to-user level has experienced similar process of change to those that happened in the period from Qin-Han to Tang-Song, featuring replacement of input quota by resource quota on a growing scale until at present when it has been extended to farm households (measurement on the farm household basis) in some irrigation areas. However, the technical and economic conditions of modern industrial civilization have gone beyond compare with ancient agricultural society. Water rights allocation is now facing new problems never been encountered before, and demand has been growing for establishing allocation mechanism at all levels above the group level. In the 25 years since China began economic reform, with the expansion in the scope of water scarcity, allocation systems have been established at all levels of the water rights structure. There are volumetric allocation systems among local governments; among groups there are the water withdrawal license systems; and there are planned water allocation systems among agricultural users. A complete set of administrative allocation systems have been established in water resource management that is marked by the enforcement of the water law revised and promulgated in 2002.

The book examines the changes in water rights structure during different periods since the founding of New China, taking the Yellow River basin as an example. During the period of planned economy, the basic form was similar to that in the Tang-Song time. During the transitional period, although the allocation of irrigation water rights remained by and large the same as in the Tang-Song period, the changes in water rights structure at other levels were completely different from ancient times. There have been administrative allocation systems among regions and groups. What merits attention is that all the levels above group in the Yellow River basin have adopted the resource quota entitlement system, that is, direct water allocation. This is, to a large extent, the result of progress in modern hydrology and information technology. Empirical study of the Yellow River basin also shows the differences between the upper and lower reaches of the river, including differences in decision-making entities and allocation levels. The differences in the irrigation management system between the upper and lower reaches of the river lie not only in the speed of institutional changes but also in the paths of institutional options. The study of these mutually influencing factors behind these phenomena reveals the non-equilibrium and path dependency of institutional changes that are determined by the social background (Coggan et al., 2013; Garrick, McCann, & Pannell, 2013). It is necessary to note that the Yellow River leads the seven major river basins in mapping out a basin-wide water allocation scheme. It has also kept pace with the institutional regime provided in the 2002 Water Law. The water rights structure of the Yellow River does not only represent the highest level of China's water rights regime but also has charted the course for the changes in water resource management of other rivers.

In general, as the changes in the ancient water rights structure, the changes of water rights structure since the founding of New China, and the process of the gradual establishment of administrative allocation mechanism are the result of natural evolution of institutions under the framework of hierarchy governance structure. Despite the fact that the book has proved the effectiveness of the administrative means in water rights allocation, the current institutions are not so satisfactory. This is understandable, from the overall perspective of history, that the period since the founding of New China is, after all, very short, and the drastic institutional change in such a short period of time is constrained by the general economic and social conditions. So it is inevitable that there are so many defects in the institutions of the water rights structure. The examination of the characteristics of water withdrawal rights reveals the low quality of water rights, which goes along well with the current administrative water allocation system. The study of the Yellow River basin has discovered that in the 15 years after 1987, the performance of the Yellow River water rights structure has gradually improved, especially since 1998, when the efficiency of administrative control system was enhanced rapidly. This indicates that the current administrative water allocation system are being adjusted and improved.

8.2.3   Perspective of Water Rights Structural Change

It is hard to make a definite prediction for the future and that is exactly the significance of conducting academic studies, because good theoretical study may provide scientific predictions for the future. It is no doubt that China will step up its pace of defining water rights in the future due to the mounting pressure of water scarcity on a large scope. In other words, there has been common understanding of the necessity of clarifying water rights. But how? There is a diversity of views. The theoretical study of this book shows that water rights are defined marginally at multiple levels and dimensions. Based on the research achievements of this book, we have seen that there is indeed a law to go by in defining water rights and that may make it possible to predict the general trend of China's water rights definition in the future.

Firstly, at the local level, more and more river basins will come out with inter-regional water allocation schemes that will define the water rights quotas for various administrative regions. And the basin that suffers from water shortages and has serious water disputes will be the earliest to map out water allocation schemes, as the 2002 Water Law has clearly provided for the establishment of inter-administrative and regional water allocation systems and emergency water diversion schemes during severe drought. The natural run-offs of rivers are allocated at the local level basically by administrative means according to the model of basin organizations offering proposals for consultation with local governments before being submitted for approval at the superordinate organs. The inter-regional water allocation schemes will become the basis for regional total volume control. Once it is worked out, it will remain stable for a fairly long period of time. Even if the scheme needs adjustments, it would be done by administrative means, however it does not rule out the compensation mechanism or even market mechanism in the process of consultation. For newly added water volume by irrigation projects, more and more market elements will be introduced in the allocation mechanisms among different localities. The project investment will be made one of the important bases for allocating the incremental water rights. This is also an important development trend in the allocation of inter-basin water diversion.

Secondly, at the group level, the water withdrawal license system will be strengthened and improved. The system was established in the 1988 water law and it has spread nationwide since then. All places have accumulated a wealth of experience in the implementation of the system. Now the system has been further revised according to the 2002 Water Law. The water withdrawal license system has clearly granted water withdrawal rights to corporate persons or individuals. In the future, it will develop toward the following directions: first, in conjuncture with the total quantity control system, it will be made the basis for approving applications for additional water withdrawal or for reducing the quotas for water withdrawal license; second, quota-based water use management system will become an important part of the monitoring and management of water withdrawal rights. Water use quota will thus become another important criterion for reducing or adjusting water withdrawal quotas; third, with the implementation of the system that requires payment for water withdrawal rights obtained, management organs will rely more and more heavily on water resource fees that serve as an economic means for regulating the amount of water withdrawal; and fourth, monitoring water withdrawal will receive increasing attention and modern information technologies will be extensively applied in this field. The improvement of the water withdrawal license system will help raise the quality of water withdrawal rights. However, the water withdrawal rights will retain its nature as administratively licensed rights for at least a period of time to come and the quality will not reach what is required of usufructuary rights. In this process, driven by policies, water rights trading may take place, but it is limited in scope and scale. Administrative means will remain the main way for adjusting water rights.

In terms of the user level, the author studies the irrigation water rights as water rights allocation, avoiding the discussion on urban water supply, because the author views that the latter falls into the category of regulation by the water supply market, following the model of making the market as dominant means, supplemented by planning, while the former follows the model of making planning a dominant means, supplemented by the market. In the future, with the progress of irrigation management reform and water pricing reform, the differences between irrigation water allocation and urban water supply will gradually be narrowed. Irrigation water management in most places is unlikely to be brought onto the path of raising the quality of irrigation water rights. Instead, it will develop toward the direction of water supply management with water measurement conducted on the household basis. This means that the water supply market model will be more and more adopted in irrigation water allocation. Such change is a deviation from the evolutionary path of the ancient Tang-Song to the Ming-Qing period. This is a very important trend of development in the modern water rights structure, as the contemporary society does not have the institutional environment on which water right trading depends as in the Qing Dynasty. In addition, the cooperation costs are very high due to weak collective actions and the costs of water measurement are low due to technical and economic progress. Such change in the transaction costs system determines that irrigation water management will not return to the Ming-Qing model. It will follow an entirely new modern path. The water supply market does not need clear allocation of water rights. It supplies water to users and fees collection is calculated according to the amount of water used. The reform pressure resulting from the expansion of the future market economy will be transmitted into the area of water resource allocation mainly through water price. The most significant change in water rights structure brought about by the market economy would be the gradual expansion and improvement of the water supply market (the water affairs market in the broad sense) at the group-to-user level, which will, in turn, exert a major impact on the allocation systems at levels above group.

8.3   Reform Strategy Proposition and Policy Recommendations

8.3.1   Shifting the Central Task from Engineering Construction to Institutional Building

The study reveals that the essence of water crisis of contemporary China is subject to serious instability of the water governance structure that is caused by inadequate payment for the management costs which is required in the maintenance of the hierarchy structure and the extensive absence of the government. Superficially, it seems to be a resource supply crisis, but in essence, it is a governance crisis caused by the inability of the water governance system to adapt to the increasingly complicated external environment due to its long-time shortfalls of demand and slow policy adjustments. The root of the absence of the government is the inability to pay for the existing high management costs that have been further raised by market-oriented reform for maintaining the hierarchy governance structure. Yet, the traditional governance structure cannot be discarded in a short period of time. The hierarchy water governance structure is, therefore, still necessary before the forces (non-governmental forces and market forces) other than the state has grown strong enough. This situation presents grave challenges to the water governance structure of contemporary China. This book argues that the only way out for China at present (the period of transition from the old system to the new) is to increase expenditure to cover the management costs, solve the problem of 'absence of the government' and pay the costs necessary to keep the hierarchy governance structure stable. That requires a large scaled institutional building led by the central government.

In general, China has concentrated efforts on engineering construction of irrigation projects over the past half century since the founding of New China. The country built more than 80,000 dams and more than 200,000 km of dykes, which is the largest in scale in the history of water conservancy. The tendency of 'paying too much attention to project construction, taking lightly project management and ignoring resource management' has led to excessive investment in hardware but input into software is inadequate. The result is that institutional building is ill-matched with project construction. This is an important reason why there is a serious 'absence of the government' in water management. It is reasonable to rely mainly on project engineering in the given historical conditions following the founding of New China. However, with the development of the times, excessive reliance on project engineering and extensive expansion of resource supply is far from being a sound policy for tackling the water crisis from the root. So change is essential. The 1998 extraordinary floods forced the Chinese government to increase sharply its investment in water control projects. In the succeeding five years, the total investment in water control projects was equal to the total investments of the previous 50 years. In 2002, the government started the costly project of diverting water from south China to north China. In this context, it has become even more urgent to shift the central task from water control project construction to institutional building.

The fundamental purpose of institutional building is to build a water-conserving and pollution-preventing society and ease the restrictions of water crisis on the economic development so as to realize the sustainable utilization of water resources and maintain a sustainable economic and social development. The core of institutional building is to improve the administrative means in water resource allocation and complete the institutional centralization of power in water resource management. This may be realized step by step. In the short run, it is necessary to intensify the enforcement of all the current water resource management institutions, especially the total amount control of regional water consumption, the water withdrawal license management, payment for water use and the monitoring and measuring systems; in the medium term, it is necessary to establish a unified water resource management model on a larger scale, including the integrated management of basins, the unified management of regional water affairs and thorough reform of the water supply system; in the long run, it is necessary to institutionalize and legalize the administrative water management system as required by the market economy. In this process, the overall institutional environment of China will continue to change. It is hopeful to complete the building of the socialist market economically and the democratic system politically. The market force and non-governmental self-governing forces would rise step by step to take over a certain measure of governmental role and feasible economic and social conditions would be prepared for the decentralization reform of water resource management. Without the centralization reform at the present stage, it is impossible to proceed with the effective decentralization reform in the future. The ultimate goal of reform is to modernize water resource management system and build a modern management model in which centralization and decentralization are well balanced under the institutional framework.

The reform of the water management system in the next period will be a tremendous challenge to the state. In the macro environment of market-oriented and decentralization reform, the centralization reform would be like rowing upstream and the costs are bound to be very high. But it must be done, as there is no other alternative under the current conditions. Otherwise, it would be impossible for the country to ease the water crisis; nor would it be possible to lay the foundation for realizing the transition from the traditional water governance model to a modern one. The reform will be tough and challenging. The key lies in a large scaled institutional building, which includes the effectuation of a shift from water control project construction to institutional building in the central task, a shift from excessive dependency on project construction to offering institutional incentive for saving water in water resource development, a shift from the concentration of hardware construction only to a harmonious development of both hardware and software, and a shift from giving priority to project investment to economic control, market regulation, social management and public service.

8.3.2 Nine Institutional Mechanisms of Institutional Building

The empirical study shows that the main problem with China's current water management is not law makings but enforcing and implementing laws. Institution is a whole set of interconnected and mutually reinforcing rules. The making of laws and regulations does not mean the building of institutions. If legal texts are to become part of institutions, there must be corresponding enforcement mechanisms to match. In fact, the 2002 water law has established a relatively complete institutional framework for water management. Although there are still rooms for improvement, what is crucial at present is whether or not the existing laws and regulations are effectively enforced and implemented, and whether or not supporting conditions can be created for the enforcement and implementation of current laws and regulations.

The reform of water management system requires a design and a drive force under a complete framework. The expanded classification of water rights system is provided in the study, that is, allocation system, enforcement system and sustenance system form a system of institutions for water resource management. These three categories of institutions are subdivided into nine mechanisms: initial allocation mechanism, reallocation mechanism, ad hoc regulation mechanism, monitoring mechanism, penalty mechanism, incentive mechanism, information mechanism, interest integration mechanism and assurance mechanism. Such classification has provided an institutional framework for the reform of the water management system. The institutional arrangements in Chinese water management should proceed from the following nine aspects:

(1) Initial allocation mechanism. The basis of water resource management is to make clear the water rights of water use units at all levels. The initial allocation mechanism is to allocate water use quotas to all user units at all levels. The 2002 Water Law has provided a whole complete set of administrative allocation systems which define the initial water use rights at each level. At the state level, there is the national water resources development strategy and medium- and long-term water supply and demand planning; at the basin level, there is the basin planning and trans-boundary runoff allocation; at the regional level, there is total amount control and quota-based management system; at the group level, there is the water withdrawal license management system; there is water fee collection system below the group level. At present, most of the major rivers in China have no water allocation schemes. According to the provision of the 2002 Water Law, all the river basins will have water allocation schemes and basins where there is water shortage, and water disputes are serious that map out their allocation schemes first as the basic reference for implementing total amount control. The water allocation schemes should give full consideration to water resource carrying capacity and water demand for domestic, economic and ecological purposes. The water withdrawal license system has become widespread. In the future, the total amount control and quota-based management must be enforced as the basis for verifying water withdrawal rights. Due to the annual fluctuations of water resources, there must be not only long-term water allocation schemes but also annual water allocation schemes based on the runoff prediction. In areas that are particularly short of water, water must be allocated within a year. The planned irrigation water allocation system is still in force in most areas below the group level and the water use quota system should be made the basis for allocation planning. The allocation of water needed by cities and towns should be regulated mainly by market through reasonable pricing system.

(2) Reallocation mechanism. After water is initially allocated, management departments must make timely adjustments in line with economic and social development, water availability and changes in water use. Once the national master plan of water resources and long- and medium-term supply and demand plans and basin plans are worked out, they should be strictly enforced and kept stable. The basin runoff allocation schemes must also be kept stable and serve as the basis of total amount control on regional water consumption. The adjustment of water rights among different regions should be carried out through democratic consultation between basin organizations and various regions, allowing the introduction of compensation mechanism and even certain elements of market. Water withdrawal right is the basis for groups to withdraw water. The validity period of water withdrawal rights is five years and the water quota should become the basis for re-verifying the water withdrawal rights during the five years. The adjustment within the five years will be done partly by management

organs according existing rules, with the quotas saved to be allowed for transfer. Below the group level, water quotas should be adjusted according to the water use quotas while strengthening the regulatory role of water price. Water needed by cities and towns should mainly be regulated by water price. In principle, the higher the level is, the more stable the water quotas should keep. The water quotas at the group level and above should be adjusted mainly by administrative means, and at the user levelwater price should become the main means for adjustment.

(3) Ad hoc adjustment mechanism. China's water resources vary greatly between years and within a year as droughts and floods are frequent. That requires corresponding ad hoc adjustment mechanism (including emergency management mechanism) to make up for what is lacking in the initial and reallocation mechanisms. The 2002 water law has provided for standby water regulation schemes during emergencies under drought and this also applies to inter-regional adjustment. This system should be implemented in the allocation of water at the basin level; and basin organizations and local governments should enhance their sense of water crisis management and get clear about the counter-measures in special circumstances. Apart from water allocated by localities, there should also be mechanism for adjusting water withdrawal rights in special circumstances. Water supply organs below the group level should also prepare crisis management schemes. In addition, it needs to be pointed out that in order to prevent abuse of power, the employment of ad hoc adjustment mechanism must be institutionalized and carried out according to an established procedure, making ex ante agreement on starting procedures, methods and time limit. Management organizations should not disrupt the original allocation schemes and water use plan unless in emergencies.

(4) Monitoring and control mechanism. Due to fluidity of water, water quotas allocation cannot be implemented automatically, and it requires enforcement through the monitoring system, including the formulation of annual water use plan and water allocation plan for the year, the control of natural run-off and the interference in the process of water use in terms of space and time. The 2002 Water Law has made clear about the unified water diversion institution, which should be implemented by administrative organs of local governments or basin organizations according to the water allocation scheme. The water management organizations at all levels should strictly follow the principle of unified water diversion, which is an important means to implement water quotas. The tasks of water diversion are arduous and coordination is difficult at the local and group levels, greater powers should be granted to regulation organs who implement water diversion schemes. China has just started water diversion and it has realized basin-wide unified diversion only in the Yellow, Heihe and Tarim river basins. It is a formidable task to realize such institution in more river basins and raise the level of water diversion.

(5) Penalty mechanism. An institution is not valid without a penalty system. Punishing law violating acts is a crucial factor for maintaining the effective operation of an institution. There are a lot of punitive articles in the current

water management laws and regulations, but many of them have not been executed in earnest manner. In the future, these penalty articles should be strictly enforced so as to uphold the authority of the allocation rules. At the local level, administrative punishments should be strengthened. In basins that are especially short of water resources, it is necessary to establish a level-by-level responsibility system in water diversion and a system for fixing responsibilities on administrative leaders and exposing acts that have violated the water allocation schemes or disobeyed water diversion orders through the media. At the group level, while implementing all kinds of management measures in overseeing water withdrawal, it is essential to make water resource fees as an important regulatory means and progressive water fee payment as an economic lever to supervise the management of water quotas. At the same time, it is necessary to make clear the terms of reference for examination and approval of new projects for withdrawing water and new water license applications according to the institution of total amount control. For projects that have violated the law, management organizations should mete out punishments. At levels below group, unreasonable water demand must be inhibited through economic means of water fee and progressive water price system.

(6) Incentive mechanism. The implementation of water quotas requires not only a powerful penalty mechanism but also an incentive mechanism in order to motivate water using units to abide by the allocation rules. At the local level, democratic consultation should be allowed in adjusting water rights under the water allocation scheme and corresponding rules. Economic compensation mechanism, or even a certain degree of market elements, may be introduced into water rights adjustments when the ecology and third party interests are not violated. Some basins may display the role of basin organizations and through the operation similar to water banks, establishing inter-regional water quota adjustment systems. When the total amount control on various regions is effectively enforced, water quotas may be allowed to turn over through the market within the regions. At the group level, water resources fees should be fixed reasonably according to different basins and different regions to give full scope to its regulatory role. Industries and cities may be encouraged to invest in water-saving projects in agriculture to exchange for saved water quotas. At levels below group, the emphasis of reform should be put on water pricing reform in agriculture, cities and towns, in order to give full scope to its regulatory role.

(7) Information mechanism. Incomplete information and information asymmetry are the important sources of costs of institutional operation. Minimizing incompleteness and asymmetry of information is an important guarantee for improving efficiency in management. Modern hydrology and information technology have created excellent conditions for this. It is, therefore, necessary to create the conditions for the extensive application of modern technologies in basin management, to improve hydrological testing systems, set up a timely regulations system that can predict the amount of rainwater and run-offs, monitor water use, regulate reservoirs water and allocate water in different river sections.

Special attention should be given to strengthening water monitoring, improving the water monitoring network, establishing the detailed statistical system and introducing the system of water information disclosure. For water withdrawal license management, overseeing organizations may, through the Internet, provide hydrologic and water use information to water users. Besides, management organizations should adopt a variety of means to make publicity and education among the public to enhance the people's awareness of water crisis and the whole society's awareness of saving water and spreading knowledge about the efficient use of water among water users.

(8) Integration mechanism. An institution is a self-sustaining system sharing common belief. Extensive identity with the institution provides a solid base for its implementation. It is considered as the biggest challenge for contemporary China of its water resources management, which is how to integrate the conflicts of interests among water-related interests groups and cultivate their common belief in the water allocation system. The most significant features of the future modern water management model in contemporary China would be an extensive introduction of a variety of mechanisms for settling the conflicts of interests under the framework of institutionalized concentration of power, including consultation mechanism, participation mechanism, interests expression mechanism and common understanding mechanism. The basin level may, in the near terms, take the lead in introducing the consultation mechanism in mediating conflicts of interests in the use of water and in the long term, make major adjustments in the basin management model by establishing basin management councils with the participation of representatives of various regions and water users as the supreme power organ in basin management, under which there may be sub-committees and advisory committees. The current basin organization may serve as an executive organ to be responsible for executing major policy decisions and maintaining routine operations. At the group level, associations may be set up as channels for water users to voice their interests and as a bridge that links up water users and management organizations. In irrigation water management, farmers should be encouraged to form water user associations as self-governing organizations at the grassroots who are involved in participation in the management of irrigation water.

(9) Assurance mechanism. Water resource management is a very complicated system that requires specialized organizations, research and financial resources for its normal operation. In order to establish and improve the above mechanisms, it is, first of all, necessary to strengthen capacity building of water management organizations, improve the human capital of water managers, make greater efforts to institutionalize and standardize management organs, making them operate according to established procedures and improving their levels of controlling and managing water by law. Secondly, it is necessary to strengthen research, increase investment in R&D, reform the research system, improve the quality of achievements and enhance the role of research system in sustaining water management. Besides, it is necessary to increase funds for water management, especially the funds for administrative enforcement of law,

regulating water, collecting, sorting out, compiling and releasing information. The raising of funds for water management should be made institutionalized and transparent so as to stamp out and prevent corruption.

No part of the above nine of mechanisms is dispensable in both the improvement of the water management system and in raising the state water management capabilities. In the next step, efforts should be made in institutional building of a well-balanced way in all aspects of water management within the above framework, making the mechanisms well adapted and well matched with one another to raise the performance of the institutional system as a whole. At present, the urgent demand is for the improvement of weak mechanisms as monitoring mechanism, incentive mechanism, information mechanism and interests integration mechanism. If China is to bring these mechanisms to perfection in major river basins in the future, it will lay an institutional foundation for the water governance to transit from the traditional model to a modern model. But it is also necessary to point out that the improvement of the above mechanisms is only a reform within the institutions. The progress of the reform of the water resource management system depends, to a large extent, on the extra-institutional reforms, including the reform of the property rights of irrigation projects, the reform of the investment and financing systems in water resource development, reform of the administrative management system and village self-govern organization reform. Institutional building involves so many aspects that it needs the society to pay very high costs or huge reform costs. In this perspective, it is really a tremendous challenge for the state governance.

8.4 Concluding Remarks: Challenge to Water Governance Is the Challenge to State Governance

This study uses the methodology of new institutional economics to study water rights problem. As water right is, in essence, an institutional option, transaction costs are the key to understanding water rights. According to the Coase Theorem, if there are zero transaction costs, the efficient outcome will occur, regardless of legal entitlement (Gjerdingen, 2014). However, in an area where the transaction costs are very high, the allocation of property rights is no longer a problem of irrelevance, instead, it is of extremely importance. Water is such a resource that it has very high transaction costs. That is why this book has conducted research with 'water rights structure' as the main thread.

Where is China's water rights structure option originated? This book traces back to 2000 years ago when a unified political civilization took shape. In the early period when productivity level was very low, natural geography was no doubt decisive in shaping up the Chinese civilization. In the ancient society that was based on an agricultural economy, when floods and drought were very frequent and the threat of floods in the lower reaches of the Yellow River was very big, the costs of social equal cooperation were extremely high. So the high degree of centralization of power by the central government to control all public affairs was an institutional arrangement as the most economical in transaction costs under all kinds of constraints.

Such form of society is hierarchical in governance structure. The logic of its construction is to use vertical administrative control to replace horizontal political transaction. This is an extension of the Coase Theorem to the area of state governance. Explaining why there is firm, Coase states that a firm replaced voluntary market transaction with enforced administrative orders. The author interprets the emergence of a unified empire as a product of replacing political transaction with administrative control. The option for the hierarchical governance structure is for the purpose of economizing on cooperation costs brought about by political transaction, however at the same time, other costs have to be paid, which illustrates that management costs is required of administrative control. So, under the hierarchical governance structure, it goes without saying that the ruler must employ all possible means to lower management costs. This led to a series of institutional arrangements that intensified the authority of the state (Gjerdingen, 2014).

These are the bases of the water rights hierarchy structure. In fact, more than 2000 years ago, it was exactly the demand for a hierarchical water governance structure that directly led to the demand for hierarchy state governance structure. That was the inherent driving force behind the unification of China by the Qin Dynasty after the country was torn by wars for more than 500 years. It is exactly because water governance has such a strong explanatory power for state governance that Guan Zi of the Spring and Autumn period said that "to govern a state, it is essential to get water under control first." Western scholars define the Chinese civilization as a hydraulic society. In the time-honored history of China, the hierarchical state governance and hierarchical water governance were reinforcing each other, mutually adapting to each other, and that is why the hierarchical governance structure was destroyed again and again and then rose again and again on the ruins of war through the changing of dynasties in an infinite cycle of upheaval and peace. The gradual change of water rights system under the framework of hierarchical structure demonstrates different characteristics during different historical periods due to technical progress.

The tide of globalization has brought about great changes to modern and contemporary China, which have never been seen over the previous thousands of years. Starting a century ago when western theories were introduced into China, changes have swept in breadth and into depth in unprecedented dimensions, especially over the past quarter of a century when China has swung its doors wide open and carried out sweeping economic and political structure reforms. Although China's water rights structure has undergone tremendous changes against the background of such economic and social transition and the concomitant water shortages, it has not deviated from the path of intensifying the hierarchy structure. The already high costs of maintaining the stability of such hierarchy structure would have become even higher as it goes along the opposite direction to the market-oriented and power sharing reforms. However, in the drastic social changes, the payment for such management costs by the state is far from enough, resulting in the absence of the government in extensive areas. The contemporary water crisis brought about by the instability of the water rights structure is, in essence, a governance crisis, due to the impotence of collective action. All these have thrown the water management system into a dilemma that has never seen before.

Where does the water rights structural change lead to in contemporary China? This is a very difficult problem and also the starting point and focus of contention in the great debate on water rights and water market. All the analyses in the book point to the correctness of this conclusion: the traditional water rights system featuring concentration of power will continue and is unlikely to be brought onto the path of decentralization in a short time. The implication of this conclusion is that water rights and water market perhaps would be the orientation of development in the long run but they have very limited roles to play in coping with the immediate water crisis. The only way out for water management during the transitional period is to carry out reforms against the tide and it concentrates efforts on solving the problem of absence of the government covering fully the costs of maintaining the hierarchy structure.

The conclusion may be surprising, but it is logically self-consistent. There may be both serious government failure and market failure in the allocation of water resources, as both administrative allocation and market allocation are very costly in transaction. The most realistic option is to compare the relative effectiveness of the two ways. The government-led water allocation that has prevailed for thousands of years, which has been proved as an effective solution. The established management model has been challenged only when there is extensive absence of government due to the transition from the planned economy to a modern market economy. However, here comes the problem: will the market become really effective once it is introduced, even if absence of the government or government failure exist? In fact, due to the absence of a collective action model that can replace traditional model of government control at the present stage of development, market failure would have more serious consequences than government failure has, if the government cannot display its role to the full in water resource allocation.

With regard to the allocation of other economic resources in China, with the progress of market-oriented reform, market has already exhibited more and more its effectiveness due to the rapidly lowering of transaction costs and the costs of direct allocation of the resources by the government are rapidly rising; so is the allocation of water resources. What is different from other economic resources is that water is characterized by a strong public character, uncertainty and strong ecological attributes. Even in the present stage when the effectiveness of market in most resource allocation areas is beyond doubt, the costs of market allocation is still far higher than administrative allocation, and that is why the market means is lagging far behind other means. If the market allocation costs decrease to the level lower than that by administrative means, water resource would be treated more as an economic asset. But this will not take place in a short period of time.

The above logic determines that our pragmatic strategic option in the reform of the water management system at the present stage is only to raise greatly the coverage of management costs by undertaking a large scaled institutional building.

The purpose of reform is to realize institutional centralization of power rather than centralization in the traditional sense; and the institutional centralization of power is for the purpose of achieving decentralization of power and ultimately a balance between centralization and decentralization. This process of transformation will not go ahead spontaneously with the expansion of the market as the market-oriented reform. On the contrary, it will meet with great resistance from groups of immediate interests and it is apt to stall, relapse or even fail on the verge of success. It is, in essence, a reform of the political system. Tens of billions or even hundreds of billions yuan of investment in public water projects may ease the crisis to a certain extent for the time being, but it can never replace the reform per se.

Can China successfully push ahead such reform and complete the transition from the traditional system to a modern system? This is a grave challenge to China in its water governance. Moreover, it is a challenge to China as to whether or not it can develop a new collective action model. It is, in essence, a challenge to the state as to whether or not it can successfully develop a modern system in state governance. The fundamental hallmark of modern state governance is that there should be a written constitution that has the de facto biggest binding force and provides a framework for clearly defining the rights and obligations of various regions, all interest groups and individuals. With this framework, the state will realize institutional centralization and decentralization of power and no longer hold residual control power without restraints; and the public and market will become a force that serves as counterparts with the state power. Without the successful modernization of state governance, it is impossible for China to modernize its water governance model.

When viewed from the angle of water management, the hydraulic society that has been maintained for thousands of years is disintegrating but far from being dying out. A new model of civilization is brewing, but it is still uncertain whether or not it can grow without a hitch. As seen from the perspective of macro history, few people could exert any action, even the smallest action, on the course of advance of history. But in the crucial period of transition from the old to the new one, the action by a key figure may have a decisive influence on the course of history. China is now for such a crucial period in the transition of the centuries-old civilization. China's successful transformation and reshaping of its traditional civilization and in the end marching toward modernization depends perhaps on the results of the games between this generation or the next and the inertia of civilization. But the more crucial aspect would be the role of state leaders in the new century, and their strong sense of historical mission, political will and their determination to carry into depth the reforms, especially reform of the political system, to push ahead successfully institutional arrangements for a modern state. I wish to conclude this book with this remark: we are undergoing the tests in this crucial period of reform; and this is true in water governance and even more so in state governance.

References

Agrawal, A. (2007). Forests, governance, and sustainability: Common property theory and its contributions. *International Journal of the Commons, 1*(1), 111–136.

Alchian, A. A., & Demsetz, H. (1972). Production, information costs, and economic organization. *The American Economic Review, 62*(5), 777.

Anderson, T. L., & Hill, P. J. (1975). The evolution of property rights: A study of the American West. *The Journal of Law and Economics, 18*(1), 163–179. doi:10.1086/466809.

Arrow, K., Sen, A., & Suzumura, K. (2011). *Chapter thirteen – Kenneth Arrow on social choice theory*. Elsevier B.V. *Handbook of Social Choice and Welfare, 2*, 3–27.

Aubin, D., & Varone, F. (2013). Getting access to water: Property rights or public policy strategies? *Environment and Planning C: Government and Policy, 31*(1), 154–167. doi:10.1068/c11247.

Barzel, Y. (1989). *Economics analysis of property rights*. Cambridge: Cambridge University Press.

Barzel, Y. (1997). *Additional property rights applications*. Cambridge: Cambridge University Press.

Barzel, Y. (2000). Property rights and the evolution of the state. *Economics of Governance, 1*(1), 25–51.

Beckmann, V., & Padmanabhan, M. (2009). *Institutions and sustainability: Political economy of agriculture and the environment – Essays in honour of Konrad Hagedorn*. Dordrecht: Springer Netherlands.

Bennett, A., & Elman, C. (2006). Complex causal relations and case study methods: The example of path dependence. *Political Analysis, 14*(3), 250–267. doi:10.1093/pan/mpj020.

Binder, S. A., Rhodes, R. A. W., Rockman, B. A., & Jessop, B. (2008). *The state and state-building*. Oxford: Oxford University Press.

Boxer, B. (2001). Contradictions and challenges in China's water policy development. *Water International, 26*(3), 335–341. doi:10.1080/02508060108686925.

Boxer, B. (2002). Global water management dilemmas: Lessons from China. *Resources, 146*, 5–9.

Bromley, D. W. (1989). Property relations and economic development: The other land reform. *World Development, 17*(6), 867–877. doi:10.1016/0305-750X(89)90008-9.

Bromley, D. W. (1995). Property rights and natural resource damage assessments. *Ecological Economics, 14*(2), 129–135. doi:10.1016/0921-8009(95)00027-7.

Bromley, D., & Pearce, D. (1992). Environment and economy – Property rights and public policy. In D. Bromley & D. Pearce (Eds.), (vol. 102, pp. 985–985). New York: Basil Blackwell.

Bromley, D. W., & Segerson, K. (1992). *The social response to environmental risk: Policy formulation in an age of uncertainty*. Boston: Kluwer Academic.

© Springer Nature Singapore Pte Ltd. 2018
Y. Wang, *Assessing Water Rights in China*, Water Resources Development and Management, DOI 10.1007/978-981-10-5083-1

Cai, S. (2001). *Scope, principles and conditions of water rights transfer* (in Chinese).

Calow, R. C., Howarth, S. E., & Wang, J. (2009). Irrigation development and water rights reform in China. *International Journal of Water Resources Development, 25*(2), 227–248. doi:10.1080/07900620902868653.

Challen, R. (2000). *Institutions, transaction costs and environmental policy: Institutional reform for water resources.* Cheltenham: Edward Elgar.

Chang, Y. (2001). *Studies of the interrupted flow of the yellow river and the water rights system.* Beijing: China Social Sciences Press (in Chinese).

Chen, X. (2002). Basin organizations and unified management of yellow river water resources. *China Water Resources, 10*, 93.

Chen, S., Wang, Y., & Zhu, T. (2014). Exploring China's farmer-level water-saving mechanisms: Analysis of an experiment conducted in Taocheng district, Hebei Province. *Water, 6*(3), 547.

Cheung, S. N. S. (1983). The contractual nature of the firm. *The Journal of Law and Economics, 26* (1), 1–21.

Cheung, S. N. S. (2000). *Economic explanation: Selected papers by Cheung* (Vol. 26). Beijing: Commercial Press.

Chi, C. A.-T. (1936). *Key economic areas in Chinese history as revealed in the development of public works for water-control.* Crows Nest: G. Allen & Unwin.

CIC, C. I. C. (Ed.) (2005). *History of irrigation and flood control in China.* Beijing: China WaterPower Press.

Coase, R. H. (1937). The nature of the firm. *Economica, 4*(16), 386–405. doi:10.2307/2626876.

Coase, R. H. (1960). The problem of social cost. *The Journal of Law & Economics, 3*, 1–44.

Coggan, A., Buitelaar, E., Whitten, S., & Bennett, J. (2013). Factors that influence transaction costs in development offsets: Who bears what and why? *Ecological Economics, 88*, 222–231. doi:10.1016/j.ecolecon.2012.12.007.

Commons, J. R. (1983). *New institutional economics (Chinese translation, Vol. 1).* Beijing: Commercial Press.

Costa, L. W. (2015). An endogenous growth model for the evolution of water rights systems. *Agricultural Economics, 46*(5), 677–687. doi:10.1111/agec.12163.

Cui, J. (2001). *Reflection on water rights, the formulation of property right law and civil law theories* (in Chinese).

Curtis, M. (2009). *Karl Marx: The Asiatic Mode of Production and Oriental Despotism.* Cambridge: Cambridge University Press.

David, P. A. (2007). Path dependence: A foundational concept for historical social science. *Cliometrica, 1*(2), 91–114. doi:10.1007/s11698-006-0005-x.

Demsetz, H. (1967). Toward a theory of property rights. *The American Economic Review, 57*(2), 347.

Demsetz, H. (1988). *Ownership, control, and the firm – The organization of economic activity.* New York: Oxford University Press.

Dou, M., & Wang, Y. (2017). The construction of a water rights system in China that is suited to the strictest water resources management system. *Water Science and Technology: Water Supply, 17*(1), 238–245. doi:10.2166/ws.2016.130.

Dukhovny, V. A., & Ziganshina, D. (2011). Ways to improve water governance. *Irrigation and Drainage, 60*(5), 569–578. doi:10.1002/ird.604.

Edelenbos, J., Meerkerk, V. I., & Van Leeuwen, C. (2015). Vitality of complex water governance systems: Condition and evolution. *Journal of Environmental Policy & Planning, 17*(2), 237–261. doi:10.1080/1523908x.2014.936584.

Eggertsson, T. (1990). *Economic behavior and institutions.* Cambridge: Cambridge University Press.

Eggertsson, T. (1997). The old theory of economic policy and the new institutionalism. *World Development, 25*(8), 1187–1203. doi:10.1016/S0305-750X(97)00037-5.

Eggertsson, T. (2009). Knowledge and the theory of institutional change. *Journal of Institutional Economics, 5*(2), 137–150. doi:10.1017/S1744137409001271.

Eggertsson, T. (2013). Quick guide to new institutional economics. *Journal of Comparative Economics*. doi:10.1016/j.jce.2013.01.002.

Eggertsson, T. (2014). Governing the commons: Future directions for the Ostrom project. *Journal of Bioeconomics, 16*(1), 45–51. doi:10.1007/s10818-013-9167-3.

Emel, J. L., & Brooks, E. (1988). Changes in form and function of property rights institutions under threatened resource scarcity. *Annals of the Association of American Geographers, 78*(2), 241–252.

Francisco, C. S., & Jorge, M. P. (2011). Risk of collective failure provides an escape from the tragedy of the commons. *Proceedings of the National Academy of Sciences, 108*(26), 10421. doi:10.1073/pnas.1015648108.

Fu, G. (1988). *Economic history of Chinese feudal society*. Beijing: People's Publishing House (in Chinese).

Furubotn, E. G. (2005). *Institutions and economic theory the contribution of the new institutional economics* (2nd ed.). Ann Arbor: University of Michigan Press.

Furubotn, E., & Pejovich, S. (1972). Property rights and economic theory: A survey of recent literature. *Journal of Economic Literature, 10*, 1137–1162.

Gao, E. (2006). *Water rights system development in China*. Beijing: China Water and Hydropower Publishing (in Chinese).

Garrick, D., McCann, L., & Pannell, D. J. (2013). Transaction costs and environmental policy: Taking stock, looking forward. *Ecological Economics, 88*, 182–184. doi:10.1016/j.ecolecon.2012.12.022.

Gjerdingen, D. H. (2014). *The normative lessons of the Coase Theorem*. World Ebook Library.

Gregg, A. (2016). After oriental despotism: Eurasian growth in a global perspective. *The Journal of Economic History, 76*, 961–963.

Grossman, S., & Hart, O. (1986). The costs and benefits of ownership: A theory of vertical and lateral integration. *The Journal of Political Economy, 94*(4), 691.

Gu, H. (Ed.) (1997). *History of water governance in China*. Beijing: China Water Conservancy and Power Publishing House (in Chinese).

Hang, Z., Zhongjing, W., You, L., & Calow, R. C. (2009). A water rights constitution for Hangjin irrigation district, Inner Mongolia, China. *International Journal of Water Resources Development, 25*(2), 373–387. doi:10.1080/07900620902868877.

Hardin, G. (1968). The tragedy of the commons. *Science, 162*(3859), 1243–1248.

Hart, B. T. (2016a). The Australian Murray–Darling basin plan: Challenges in its implementation (Part 1). *International Journal of Water Resources Development, 32*(6), 819–834. doi:10.1080/07900627.2015.1083847.

Hart, B. T. (2016b). The Australian Murray–Darling basin plan: Challenges in its implementation (Part 2). *International Journal of Water Resources Development, 32*(6), 835–852. doi:10.1080/07900627.2015.1084494.

Hu, A. (Ed.) (2003). *Report on national conditions that influence policy decision making*. Beijing: Tsinghua University Press (in Chinese).

Hu, A., & Wang, Y. (2000). Public policy of water resources allocation in the transition: Quasi-market and democratic and consultative politics. *Chinese Soft Science, 5*, 5–11 (in Chinese).

Hu, Z., Fu, C., & Wang, X. (2003). *Allocation and management of water property right*. Beijing: Science Press.

Huang, S. (1995). *Introduction to economics of property rights*. Jinan: Shangdong People Press (in Chinese).

Huang, R. (2002). *On Chinese history by the Hudson river*. Shanghai: Sanlian Bookstore (in Chinese).

Israel, M., & Lund, J. R. (1995). Recent California water transfers: Implications for water management. *Natural Resources Journal, 35*(1), 1–32.

Jiang, W. (2001). *Studies of the basic water rights theories*. Water rights and water market (Second Series of Selected materials, Volume II).

Jin, L., Zhang, G., & Tian, H. (2014). Current state of sewage treatment in China. *Water Research, 66*, 85.

Kauffman, G. (2015). Governance, policy, and economics of intergovernmental river basin management. *An International Journal – Published for the European Water Resources Association (EWRA), 29*(15), 5689–5712. doi:10.1007/s11269-015-1141-5.

Klein, B. (1983). Contracting costs and residual claims: The separation of ownership and control. *The Journal of Law and Economics, 26*(2), 367–374. doi:10.1086/467040.

Krutilla, K., & Krause, R. (2010). Transaction costs and environmental policy: An assessment framework and literature review. *International Review of Environmental and Resource Economics, 4*, 261–354.

Lake, D. A. (1996). Anarchy, hierarchy, and the variety of international relations. *International Organization, 50*(1), 1–33.

Lang, M. (2011). Globalization and global history in Toynbee. *Journal of World History, 22*(4), 747–783.

Libecap, G. D. (1986). Property rights in economic history: Implications for research. *Explorations in Economic History, 23*(3), 227–252. doi:10.1016/0014-4983(86)90004-5.

Lin, Y. (1994). *Induced change and coercive change: On the economic theory about institutional change*. Shanghai: Shanghai Sianlian Publishing House (in Chinese).

Lin, J. Y. (2013). New structural economics: The third wave of development thinking. *Asian-Pacific Economic Literature, 27*(2), 1–13. doi:10.1111/apel.12044.

Liu, B. (2002, 18 May). *An analysis of China's water rights system.* Paper presented at the Comprehensive Thesis Training of Tsinghua University, Beijing, Tsinghua University (in Chinese).

Liu, J., Zang, C., Tian, S., Liu, J., Yang, H., Jia, S., et al. (2013). Water conservancy projects in China: Achievements, challenges and way forward. *Global Environmental Change, 23*(3), 633–643. doi:10.1016/j.gloenvcha.2013.02.002.

Liu, F., Duan, Y., & Deng, Y. (2014a). Pricing mechanism for water right trading in China. *China Water Resources, 20*, 1–3.

Liu, Q., Wang, B., & Chen, G. (2014b). Water right system in China and its relationship with water management. *China Water Resources, 20*, 4–6.

Lukasiewicz, A., & Dare, M. (2016). When private water rights become a public asset: Stakeholder perspectives on the fairness of environmental water management. *Journal of Hydrology, 536*, 183–191. doi:10.1016/j.jhydrol.2016.02.045.

Ma, J. (2003). Transaction cost political science: Current conditions and prospects. *Economic Studies, 1*, 80–86 (in Chinese).

Mao, A. (2002). Deepen reform to promote sustainable development in Weishan irrigation area. *China Water Resources, 5*, 41–59.

Marshall, G. R. (2013). Transaction costs, collective action and adaptation in managing complex social–ecological systems. *Ecological Economics, 88*, 185–194. doi:10.1016/j.ecolecon.2012.12.030.

Marx, K., & Engels, F. (1972). Chinese 4 – Volume "Selected Works of Marx and Engels" Published. *Peking Review, 15*(19), 3.

McCann, L. (2004). Induced institutional innovation and transaction costs: The case of the Australian National Native Title Tribunal. *Review of Social Economy, 62*(1), 67–82. doi:10.1080/0034676042000183835.

McCann, L. (2013). Transaction costs and environmental policy design. *Ecological Economics, 88*, 253–262. doi:10.1016/j.ecolecon.2012.12.012.

McCann, L., Colby, B., Easter, K. W., Kasterine, A., & Kuperan, K. V. (2005). Transaction cost measurement for evaluating environmental policies. *Ecological Economics, 52*(4), 527–542. doi:10.1016/j.ecolecon.2004.08.002.

McDaniel, B. A. (2001). The crisis in social and institutional integration. *The Social Science Journal, 38*(2), 263–275. doi:10.1016/S0362-3319(01)00112-4.

Mcginnis, M. (2000). *Polycentric governance and development (translated into Chinese by Shoulong Mao)*. Shanghai: Sanlian Bookstore.

Meinzen-Dick, R. S., Brown, L. R., Feldstein, H. S., & Quisumbing, A. R. (1997). Gender, property rights, and natural resources. *World Development, 25*(8), 1303–1315. doi:10.1016/S0305-750X(97)00027-2.

MEPPRC. (2016). Annual report on the State of Water Environment in China. Retrieved from http://jcs.mep.gov.cn/hjzl/zkgb/2009hjzkgb/201006/t20100603_190435.htm (in Chinese)

Miller, G. J. (1992). *Managerial Dilemmas: The political economy of hierarchy*. Cambridge: Cambridge University Press.

Mitchell, B., Priddle, C., Shrubsole, D., Veale, B., & Walters, D. (2014). Integrated water resource management: Lessons from conservation authorities in Ontario, Canada. *International Journal of Water Resources Development, 30*(3), 460–474. doi:10.1080/07900627.2013.876328.

Moe, T. (1984). The new economics of organization. *American Journal of Political Science, 28*, 739–777.

Murray, T. (2016). *Contesting property rights*. Cambridge: Cambridge University Press.

Myint, H., & Cheung, S. N. S. (1970). Theory of Share Tenancy. *Economica, 37*(147), 329. doi:10.2307/2551987.

Needham, J. (1981). *Science in traditional China: A comparative perspective*. Cambridge: Harvard University Press.

North, D. C. (1981). *Structure and change in economic history*. New York: W. W. Norton.

North, D. C. (1990). *Institutions, institutional change, and economic performance*. Cambridge: Cambridge University Press.

North, D. C., & Thomas, R. P. (1973). *The rise of the Western World: A new economic history*. Cambridge: Cambridge University Press.

Olson, M. (1971). *The logic of collective action: Public goods and theory of groups*. Cambridge: Harvard University Press.

Olson, M. (1982). *The rise and decline of nations: Economic growth, stagflation, and social rigidities*. New Haven: Yale University Press.

Olson, M. (1990). Economy, logic, and action. *Society, 27*(3), 71–81. doi:10.1007/BF02695542.

Olson, M. (1996). Big bills left on the sidewalk: Why some nations are rich, and others poor. *The Journal of Economic Perspectives, 10*(2), 3.

Ostrom, E. (1990). *Governing the commons: The evolution of institutions for collective action*. Cambridge: Cambridge University Press.

Perdue, P. (1990). Lakes of Empire: "Man and Water in Chinese History". *Modern China, 16*(1), 119.

Perman, R., Ma, Y., & McGilvray, J. (1996). *Natural resource and environmental economics*. London and New York: Longman.

Philpot, S., Hipel, K., & Johnson, P. (2016). Strategic analysis of a water rights conflict in the south western United States. *Journal of Environmental Management, 180*, 247–256. doi:10.1016/j.jenvman.2016.05.027.

Pietz, D. (2010). Researching the state and engineering on the North China Plain, 1949–1999. *Water History, 2*(1), 53–60. doi:10.1007/s12685-010-0017-0.

PLDMWR, P. a. L. D. o. t. M. o. W. R. (Ed.) (2002a). *Materials for the study of water law*. Beijing: Water Conservancy and Hydropower Publishing House (in Chinese).

PLDMWR, P. a. L. D. o. t. M. o. W. R. (Ed.) (2002b). *Speeches on the Law of the People's Republic of China*. Beijing: China Water Conservancy and Hydropower Publishing House.

Putterman, L., & Kroszner, R. S. (Eds.). (2000). *Economic nature of firms (translated into Chinese by Sun Jingwei)*. Shanghai: Shanghai Financial and Economic Press.

RIDMWR, R. I. D. o. t. M. o. W. R. (1999). Outlined History of Farmland Irrigation in New China (1949–1998). Beijing: China Water Resource and Hydropower Publishing House.

Rymes, T. K., & Gordon, S. (Eds.). (1991). *Welfare, property rights and economic policy essays and tributes in honour of H. Scott Gordon*. Ottawa: Carleton University Press.

Schlager, E., & Ostrom, E. (1993). Property-rights regimes and coastal fisheries: An empirical analysis. In T. L. Aderson & R. T. Simmons (Eds.), *The political economy of customs and culture: Informal solutions to the commons problem* (pp. 13–42). Lanham: Rowman and Littlefield.

Schmidt, J. J., & Mitchell, K. R. (2014). Property and the right to water. *Review of Radical Political Economics, 46*(1), 54–69. doi:10.1177/0486613413488069.

Schmidt, J., & Shrubsole, D. (2013). Modern water ethics: Implications for shared governance. *Environmental Values, 22*(3), 359–379. doi:10.3197/096327113X13648087563746.

Schroeder, N. M., & Castillo, A. (2013). Collective action in the management of a tropical dry forest ecosystem: Effects of Mexico's property rights regime. *Environmental Management, 51* (4), 850–861. doi:10.1007/s00267-012-9980-9.

Shen, D. (2004). The 2002 Water Law: Its impacts on river basin management in China. *Water Policy, 6*(4), 345–364.

Shen, D. (2014). Post-1980 water policy in China. *International Journal of Water Resources Development*, 1–14. doi:10.1080/07900627.2014.909310.

Sheng, H. (2002). *Discussions on the theories of water rights system and water market.* Retrieved from Unirule Economics Research Institute, Beijing (in Chinese).

Sima, Q. (2007). *The First Emperor: Selections from the Historical records.* Oxford: Oxford University Press.

Simpson, L., & Ringskog, K. (1997). *Water markets in the Americas.* Washington, DC: The World Bank.

Song, S. (2016). Governance, efficiency, and development. *The Chinese economy, 49*(2), 57–59. doi:10.1080/10971475.2016.1142822.

Speed, R. (2009a). A comparison of water rights systems in China and Australia. *International Journal of Water Resources Development, 25*(2), 389–405. doi:10.1080/07900620902868901.

Speed, R. (2009b). Transferring and trading water rights in the People's Republic of China. *International Journal of Water Resources Development, 25*(2), 269–281. doi:10.1080/07900620902868687.

Sun, T., Wang, J., Huang, Q., & Li, Y. (2016). Assessment of water rights and irrigation pricing reforms in Heihe River Basin in China. *Water, 8*(8), 333. doi:10.3390/w8080333.

Toynbee, A.J. (1989). *A study of history.* Hangzhou: Zhejiang People's Publishing House (in Chinese).

Twitchett, D. C., & Fairbank, J. K. (Eds.). (1978). *The Cambridge history of China.* Cambridge: Cambridge University Press.

Umbeck, J. (1981). Might makes rights: A theory of the formation and initial distribution of property rights. *Economic Inquiry, 19*(1), 38–59. doi:10.1111/j.1465-7295.1981.tb00602.x.

Van de Ven, A. H., & Lifschitz, A. (2009). John R. Commons. In P. Adler (Ed.), The Oxford handbook of sociology and organization studies, classical foundations Oxford: Oxford University Press.

Waley, A. (2012). *Three ways of thought in ancient China: China: History, philosophy, economics 36.* United Kingdom: Routledge.

Wang, S. (2000). *Water rights and water market: Economic means for optimizing the allocation of water resources.* Speech at the Annual Meeting of the China Water Resources Society, Beijing.

Wang, Y. (2001). China's institutional reform of water resources allocation as seen from water rights trading between Dongyang and Yiwu Cities. *China Water Resource, 6*, 35–37 (in Chinese).

Wang, S. (2002). *Resource, water and power: Man and nature in harmony.* Beijing: China Water Conservancy and Hydropower Publishing House (in Chinese).

Wang, S. (2003a). *Use the sustainable utilization of water resources to stimulate the sustainable economic and social development.* Paper presented at the Ministerial Meeting of the Third World Water Forum, Tokyo, Japan.

Wang, Y. (2003b). *How should China build a water-efficient society: Investigation report on the experiments in building a water-efficient society in Zhangye, Gansu Province*. Paper presented at the CAS-Tsinghua Center for China Studies, Beijing.

Wang, Y. (2003c). Water dispute in the Yellow River Basin: Challenges to a centralized system. *China Environment Series, 3*, 94–98.

Wang, Y. (2005). *Economic explanation of China's water rights*. Shanghai: Shanghai People's Publishing House (in Chinese).

Wang, Y. (2013a). *Research on the stages of water conservancy development in China*. Beijing: Tsinghua University Press (in Chinese).

Wang, Y. (2013b). *Water governance reform in China*. Beijing: Tsinghua University Press (in Chinese).

Wang, Y., & Hu, A. (2002). Water resources management mode in the Yellow River basin should be better governance than control. *Yellow River, 1*, 23–25 (in Chinese).

Wang, Y., & Hu, A. (2007). Multiple forces driving China's economic development: A new analytic framework. *China & World Economy, 15*(3), 103–120. doi:10.1111/j.1749-124X.2007.00071.x.

Wang, Y., & Tian, F. (2010). Evaluation and prospect on the pilot program of water right transfer in the Yellow River basin. *China Water, 1*, 21–25 (in Chinese).

Wang, Z., Zhu, J., & Zheng, H. (2015). Improvement of duration-based water rights management with optimal water intake on/off events. *Water Resources Management, 29*(8), 2927–2945. doi:10.1007/s11269-015-0979-x.

WB, W. B. (2016). World Development Indicators. Retrieved 8th September 2016, from http://data.worldbank.org/data-catalog/world-development-indicators

Weber, K. (1997). Hierarchy amidst anarchy: A transaction costs approach to international security cooperation. *International Studies Quarterly, 41*(2), 321.

White, C., & Costello, C. (2011). Matching spatial property rights fisheries with scales of fish dispersal. *Ecological Applications: A Publication of the Ecological Society of America, 21*(2), 350.

Williamson, O. E. (1977). Markets and hierarchies. *Challenge, 20*(1), 70–72.

Williamson, O. E. (1985). *The economic institutions of capitalism: Firms, markets, and relational contracting*. New York: Free Press.

Williamson, O. E. (2000). The new institutional economics: Taking stock, looking ahead. *Journal of Economic Literature, 38*(3), 595–613.

Wittfogel, K. (1957). *Oriental Despotism: A comparative study of total power*. New Haven, CT: Yale University Press.

Wolf, A. T. (1998). Conflict and cooperation along international waterways. *Water Policy, 1*(2), 251–265. doi:10.1016/S1366-7017(98)00019-1.

Wouters, P., Hu, D., Zhang, J., Tarlock, D., & Andrews-Speed, P. (2004). The new development of water law in China. *University of Denver Water Law Review, 7*(2), 243–308.

Wu, J. (2002). *Introduction to contemporary water resources management*. Beijing: China Water Power Press.

Xia, C., & Pahl-Wostl, C. (2012). The development of water allocation management in the Yellow River basin. *Water Resources Management, 26*(12), 3395–3414. doi:10.1007/s11269-012-0078-1.

Xiao, Z. (1999). Water rights problems in farmland irrigation in the Guanzhong Area during different historical periods. *Studies of Economic History of China, 1*, 48–64.

Xu, X. (2002). Practice of price reform of water used in agriculture in the Weishan Irrigation Area. *China Water Resources, 8*, 46–62.

Yang, L. (2002). From Dongyang-Yiwu water rights transfer to understand State Property Rights in the Transition *Renmin Zhujiang*.

Yang, H., & Jia, S. (2008). Meeting the basin closure of the Yellow River in China. *International Journal of Water Resources Development, 24*(2), 265–274. doi:10.1080/07900620701723497.

Yang, H., Flower, R. J., & Thompson, J. R. (2013). Sustaining China's water resources. *Science (New York, NY), 339*(6116), 141. doi:10.1126/science.339.6116.141-b.

YRHEO, Y. R. H. E. O. (Ed.) (1995). *History of the administration of the Yellow River.* Zhengzhou: Henan People's Publishing House (in Chinese).

Zhang, W. (1996). Ownership, governance structure and commission-agent relations. *Economic Studies, 9,* 71–80 (in Chinese).

Zhang, J. (2012). *Types of hydraulic society.* Beijing: Peking University Press.

Zhao, W. (2001). *Improving the water rights system of China* (in Chinese).

Zhou, Q. (2000). Nature of publicly owned enterprises. *Economic Studies, 11,* 3–12 (in Chinese).

Zhou, Q. (2002). *Changes in proper rights and system: Empirical studies of China's reform.* Beijing: Social Sciences Documentation Press.

Zhu, L. (Ed.). (2004). *The great Yellow River.* Zhengzhou: Yellow River Conhservancy Press.

Index

A

Administration, 17, 60, 76, 107, 139, 144–146, 148

B

Bundle of water rights, 24, 65
Bureaucracy, 28, 33, 37, 73, 86

C

Case study, 18, 177
Chan quan, 22
Chinese civilization, 1, 15, 16, 40, 55, 60, 61, 80, 226, 227
Coase Theorem, 21, 28, 32, 36, 92, 226, 227
Cooperation costs, 29, 50–53, 55, 57–59, 89, 103, 105, 204, 219, 227

D

De facto property right, 22
De facto rights, 23, 27, 193, 196, 208
Decision making right, 65, 87
Decision-making entity, 69, 71, 80, 81, 86, 105
Decision-making power, 69, 71, 73, 80, 86, 87
Divisibility, 101, 122, 133, 162, 206, 209
Dongyang City, 9, 191–193
Droughts, 1, 2, 40, 41, 43, 53, 56, 60, 96, 110, 223

E

Elinor Ostrom, 14, 68, 69

Environment, 2–5, 7–9, 18–20, 31, 33, 34, 37, 38, 42, 43, 48, 51, 53, 56, 61, 67, 68, 72, 86, 87, 89, 96, 99, 104–106, 108, 126, 129–131, 139, 141, 157, 158, 179, 183, 187, 193, 201–204, 213, 215, 219–221
Equal entities, 27
Equilibrium, 50, 53, 56–58, 91–94, 98, 99, 105, 106, 210, 216
Exclusivity, 91, 101, 206, 209

F

Floods, 1–3, 5, 6, 16, 19, 20, 24, 39–43, 53–56, 60, 61, 96, 110, 113, 138, 220, 223, 226

G

Governance ability demand curve, 60–62
Government absence, 18, 176, 189, 212
Guan Zong, 41

H

Haihe River, 3, 40, 41, 194, 195
Hierarchy, 13, 14, 16–18, 28–31, 33–37, 48, 49, 51–54, 56–58, 63, 69–74, 80–82, 85, 87, 89, 95, 96, 99, 102–106, 109, 119, 123, 126–128, 135, 141, 160, 173, 176, 177, 179, 204, 207, 211–215, 217, 219, 227, 228
Hierarchy conceptual model, 16, 18, 28, 74, 80, 96, 99, 109, 123, 128, 173, 176, 177
Hierarchy governance structure, 18, 49, 81, 87, 102, 103, 123, 135, 211–213, 215, 217, 219

© Springer Nature Singapore Pte Ltd. 2018
Y. Wang, *Assessing Water Rights in China*, Water Resources Development
and Management, DOI 10.1007/978-981-10-5083-1

Printed in the United States
By Bookmasters